"Buckminster Fuller was one of those world historic geniuses who reminded us of the extraordinary things that are possible and inspires all of us to set about doing them! This book elucidates his thinking, honors his spirit and creates an enthusiasm for continuing his work."

—Marianne Williamson, author of *A Return To Love* and *Healing the Soul of America*

"Prepare to have your world turned upside down—or more correctly, really, right side up. There is enough food, energy, and resources to support all the world's population at a very high standard of living. This book of great quotations from Buckminster Fuller—the twentieth century's Leonardo da Vinci—shows us how to take evolution's next step as exemplified by the words of forty-two guest commentators who are doing just that."

—Steve Wozniak, cofounder of Apple Computer and chief scientist for Fusion-io

"Buckminster Fuller was one of the greatest teachers I have ever known. This new book, *A Fuller View*, brings his incredible teachings into full view at a time when we need to hear his clear and wise voice as never before. I am grateful that Steven Sieden has put together this brilliant piece of work that everyone alive should read. It will blow your mind in the best possible way and set you on a path of truth, transformation, and integrity. Buy it, read it, tell everyone! Bucky is back!"

—Lynne Twist, author of *The Soul of Money* and cofounder of the Pachamama Alliance

"Buckminster Fuller manifested his dreams in a way that contributed to us all. Now, Steven Sieden restates Bucky's vision so that we can all follow the path and wisdom of a great man in manifesting change and creating abundance. I recommend everyone read *A Fuller View* and recognize it as yet another message from the universe."

—Mike Dooley, author of *Notes From the Universe* (tut.com) and *Manifesting Change*

"These pages are bursting forth with wisdom, compassion, insight, and exuberant discovery. A book to be read and re-read, to be shared, absorbed, and implemented. Timely, transformative, and deeply engaging, it opens new doors into the rich and brilliant legacy of Bucky Fuller."

—Velcrow Ripper, director of *Occupy Love, Fierce Light,* and *Scared Sacred*

"If there's anything as rewarding as reading Bucky Fuller, it is opening pages of dialogue with him by some of the heartiest minds of our time. The ongoing dialogue brings ongoing life. Praise to Steve Sieden for making this happen."

—Joanna Macy, eco-philosopher, Buddhist scholar, and author of *World As Lover, World As Self* and *Widening Circles*

"Buckminster Fuller was a key guiding light for most of us still engaged around the world and striving to bring forth his vision of a just, peaceful, ecologically sustainable world. This book is one of the great literary treasures of the twenty-first century ."
  —Hazel Henderson, author of *Building a Win–Win World* and *Ethical Markets: Growing The Green Economy*

"Steven Sieden has captured the many facets of Buckminster Fuller in a way that is both understandable and inspiring. The inclusion of a wide array of guest commentators allows *A Fuller View* to provide a unique vision of Bucky's comprehensive strategy and reminds us that we're all in this together on the planet Bucky named Spaceship Earth."
  —John Robbins, author of *No Happy Cows*, *The Food Revolution*, *Diet for a New America*, and *The New Good Life*

"*A Fuller View* let me see the remarkable man from many new facets of the unique polyvertexion that is Bucky. As powerful as Fuller's insights will always be, reading the impact his presence had on so many lives helped humanize one of the great minds of my lifetime."
  —Foster Gamble, creator and host of the *THRIVE* movie

"Bucky Fuller challenged the system and offered positive solutions to all our challenges—both global and personal. *A Fuller View* reminds us that each of us can make a difference and using Bucky's solutions will benefit us all. He's one of my heroes and I'm so very proud to be included as a contributor."
  —Robert White, executive coach and author of *Living an Extraordinary Life*

"*A Fuller View* offers inspiration for how to respond creatively, compassionately, and wisely to the myriad of opportunities and challenges of these times. A brilliant collection of insights from people who have taken Bucky's wisdom to heart!"
  —Dr. Joel Levey and Michelle Levey, social architects, founders of WisdomAtWork .com, and authors of *Luminous Mind*, *Wisdom at Work*, and *Living in Balance*

"Steven Sieden has assembled a wonderful feast of wisdom and warmth from a myriad of folks who were touched and guided by Bucky Fuller's great Heart and Mind. Be generous with yourself and read this book! It could inspire you to change your life—and the condition of humanity—in unexpected ways."
  —Jack Elias, trainer and hypnotherapist with an international clientele and author of *Finding True Magic*.

SIEDEN

# A FULLER VIEW

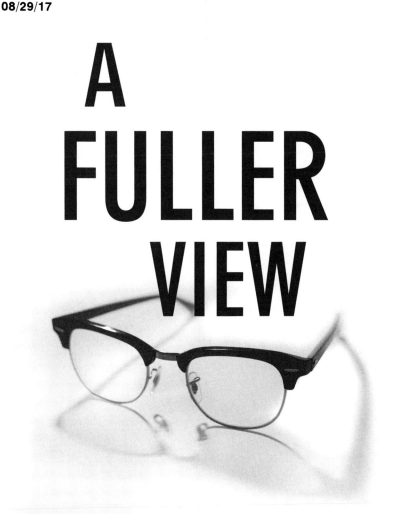

# BUCKMINSTER FULLER'S VISION
## of Hope & Abundance For All

DIVINE
ARTS

Published by DIVINE ARTS
DivineArtsMedia.com

An imprint of Michael Wiese Productions
12400 Ventura Blvd. #1111
Studio City, CA 91604
(818) 379-8799, (818) 986-3408 (FAX)

Cover design: Johnny Ink
Cover photo: © Wernher Krutein, www.photovault.com
Book Layout: William Morosi
Copyediting: Andrew Beierle
Printed by McNaughton & Gunn, Inc., Saline, Michigan

Manufactured in the United States of America
Copyright 2011 by L. Steven Sieden

Library of Congress Cataloging-in-Publication Data

A Fuller view : Buckminster Fuller's vision of hope and abundance for all / [edited by] Steven Sieden.
    p. cm.
Includes index.
ISBN 978-1-61125-009-1
1. Fuller, R. Buckminster (Richard Buckminster), 1895-1983. I. Sieden, Lloyd Steven.
TA140.F9F85 2012
620.0092--dc23
                            2011039734

DEDICATED TO ALL SENTIENT BEINGS AND THE LEGACY OF BUCKMINSTER FULLER—SPREADING SEEDS OF POSSIBILITY THAT CONTINUE TO SPROUT AND BLOOM ON BEHALF OF ALL LIFE. MAY ALL MY RELATIONS AND ALL MY CHILDREN BENEFIT FROM THIS WORK AND THE KIND COMPASSION CONTAINED IN THOSE SEEDS. AND MAY SAMARA, JACKSON, ORLI, ZACHARY, JESSE, DANIEL, KARLEY, AND DAVID ALL FIND THEIR TRUE PLACE AND PATH IN OUR EMERGING "WORLD THAT WORKS FOR EVERYONE."

# Table of Contents

# FOREWORD

## By Michael Wiese

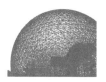 *"When we speak of the integrity of the individual, we speak of that which life has taught the individual by direct experience.... It was (this) realization that brought the author to reorganize his life to discover what, if anything, could the little, penniless, unknown individual be able to do effectively on behalf of all humanity that would be inherently impossible for the great nations or great corporate enterprises to do.*

*"Ninety-nine percent of humanity does not know that we have the option to 'make it' economically on this planet and in the Universe. We do. It can only be accomplished, however, through a design science initiative and technological revolution."*—Critical Path

I HAD THE EXTREME GOOD FORTUNE TO KNOW BUCKY, TO see him in action, and to try to put his teachings into practice. He influenced and changed the direction of my life.

## 1970

I am in my early twenties, living in Tokyo, when someone sends me Stewart Brand's *The Whole Earth Catalog*. I order all of Bucky's books and read them in my tiny *tatami* room. This futuristic city is designed to accommodate and service the needs of its teeming population and therefore seems like a good place to think about how to make the world work for everyone.

In a world convinced that *scarcity* is the norm, Bucky is the first person I've ever heard who is certain the world holds *abundance* and that by employing design science we can raise the standard of living for everyone. Sitting in a temple in Kyoto, it occurs to me that Bucky may have taken the vow of the *bodhisattva.*

## 1972

I visit my folks in Champaign, Illinois. I've been invited to a luncheon lecture at my mother's Art Club. As the red Jell-O is being passed around, the din of silverware on china is hushed by a loud noise as the speaker bangs his shoe on the podium. It is Buckminster Fuller. *"It's five minutes to midnight! This is humanity's final exam!"* Shocked silence. Inwardly, I cheer, "Go Bucky Go"! Afterwards, I hurry up to meet the great man.

## 1975

In San Francisco, I sit around the kitchen table with a friend planning a series of events to be called "Guardians of the Planet." Bucky Fuller, Jacques Cousteau, and Frederick Leboyer are on our wish list. We start with Bucky, but after hearing him, we go no further down our list.

## 1977

I arrange my life so I can work on events and films with Bucky; anything to be in his presence. During filming at Pajaro Dunes, Bucky walks along the beach. Suddenly he stops, stoops down, and scoops up a handful of white foam from the surf. *"Dear boy, you don't think Nature uses pi when she creates these magnificent bubbles?"*

## 1978

I've had the good fortune of sharing a July 12th birthday with Bucky. We were able to celebrate it together a couple of times. Once was in Bali. He invites friends and colleagues to a mini-conference he calls Campuan (the meeting of two rivers) to share global or universal experiences. Attendees include Kenneth Clarke, Werner Erhard, Lim Chong Keat, Nina Rockefeller, Arie Smit, Shirley Sharkey, and others. I attend as a "junior varsity" member and tape record the event.

At the birthday party, the Balinese surprise Bucky with a bamboo dome, blessings, flowers, and offerings, which they present in his honor. Bucky loves the Balinese. He says they exemplify natural cooperation and, like the crew of a sailing ship, spontaneously know what to do. This suggests to me a natural knowing and interconnectedness: something Bucky and the Balinese have tapped into.

When we return from Bali, a group of friends host a dinner for Bucky. After dinner, Neal Rogin and I perform a shadow play we had written for Bucky. I made a shadow puppet that in silhouette looks like him. As I recall, the story takes place in heaven as a being is about to be sent down to Earth on a mission to save the planet. Bucky sits in a lawn chair, surrounded by young people, totally transfixed by the shadow play. Then Morgan Smith sings a song she had written for him called "Great Grandfather," which expresses for all of us our deep affection for Bucky. He doesn't say anything but I can see tears well up behind his thick glasses.

One beautiful October, as the leaves change to red and yellow, I visit Bucky in Maine. He has just returned from a stint in the hospital and is anxious to get his beloved sailboat *Intuition* back into the water. We film him sailing around his house on Bear Island for an interspecies communication film called *Dolphin*.

We watch in wonder as dolphins ride the bow wave, weaving in and out below us. *"Beautiful, just beautiful. Nature wouldn't have given the dolphin such a large brain unless she was doing something quite extraordinary with it."* This statement reflects my own understanding and I spend the next couple years making films on dolphins taking this message to the world.

## 1979

As a mentor and teacher, Bucky is inexhaustible. At dinner together he fills the tablecloth with drawings of tetrahedrons and tutors me in synergetic mathematics. I tell Bucky that I just don't get it but he persists, knowing that unless we all understand how to employ Nature's principles, humanity will be doomed with its antiquated way of thinking about the Earth's resources.

## 1982

Even at 86, Bucky is still traveling around the world, speaking to an average of twenty-five hundred people a day (he keeps track of these things). He is giving a big public lecture at the Frank Lloyd Wright–designed Marin Civic Center just north of San Francisco. The event is sold out. I hadn't planned to go but at the last minute I do. It has been arranged that someone would drive Bucky back to his hotel after the lecture but the young woman who had agreed to do it drops the ball and doesn't show up because she heard that John Denver (Bucky's friend), whom she wants to meet, is not going to be there after all.

When I learn this I go backstage to see if new plans have been made. Bucky isn't in his dressing room. I find him tired, lost, and confused among theatre flats backstage. Bucky has given his all. Morgan and I take him to a restaurant in Sausalito for his favorite: steak and potatoes. When his strength returns, he makes Morgan sing "Great Grandfather" a cappella! He doesn't care that the restaurant is packed with people.

## 1983

I am with Bucky a few days before he dies a conscious death. His wife, Ann, is ill and goes into a coma. She has told Bucky she is afraid of going first and so Bucky has promised her that *he* would go first and meet her on the other side. He sits with her, holds her hand, and then dies. She dies shortly thereafter.

In Boston, I attend the double funeral ceremony in the country's oldest cemetery where I learn of Bucky's great love for Ann. He had built the great Montreal dome—his Taj Mahal—for her. Later when I tell friends about it, I keep having a kind of Freudian slip: I say, "I was in Boston at a wedding," because that's how it felt to me.

## 2011

I am no longer a young man. Without knowing exactly when it happened I've become an elder and find I have to step up to all that this entails. I think of Bucky, how he'd taken me under his wing, allowed me access, and mentored me. What an opportunity.

I meet 18-year old Simon Olszewski in a ceremony in the Peruvian Amazon. He comes from Australia and is interested in plant medicines and filmmaking. He is open and curious about everything, and ready to learn. He reminds me of me. It will be his generation that will shape our future. I invite Simon to Bali to assist on a film I am making about Balinese *taksu* (divine inspiration). We pass where Bucky's dome once stood but it is no longer there. The insects, the rain, and sun have recycled it. Everything is dynamic and changing. Bucky certainly marveled at these great natural and cosmic forces and embraced them in his synergetics.

As my daughter navigates the changes of her teenage years, I know that the "reality" that my generation sees will not be the reality she sees. Hopefully, as Bucky would have wished, it will be more in keeping with Nature's principles.

It is thrilling to see the spontaneous cooperation that arose when Steven Sieden put forth the notion of this book. The dozens of "commentaries" that you will read are testimony to

My path working with Fuller and the two Fuller books I have completed thus far did not begin with such lofty aspirations because no one had done the task of distilling Bucky's wisdom down to an essence. My journey began on Laguna Beach, CA in 1981.

I had just devoted a year of my life to a very draining but rewarding job, and I needed to replenish my depleted energy and spirit. Years before, I had been inspired by Bucky's vision and passion, and in 1981 I took my copy of his latest book, *Critical Path,* to the beach every day for three months.

In that pristine setting, I studied every page and learned as best I could. Then, I discovered that at the age of eighty-six Bucky was still active and lecturing and had a home in Southern California. I also found out that his grandson had started the "Friends of Buckminster Fuller Foundation" not far from where I was living.

On that beach I also initiated my own "experiment" to determine how many of Bucky's ideas and life strategies resonated with my personal experience and were useful to me. My experiment has now continued for over thirty years, and it has provided me with some amazing insights that I share with you throughout these pages.

I offer these insights as a tribute to Buckminster Fuller and what I have found to be true. Only after I had validated them as spot on and useful was I willing to share them with others. The fact that this book contains only insights and ideas I have found to be valid, viable, and practical is because of my personal experiment to determine and document the most vital aspects of Bucky's life and teachings.

That, however, is no reason for you to accept or believe anything written by me or our guest commentators. Do as Bucky did and check them out against your personal experience. See if they fit for you. If something works, use it. If it doesn't, let it go and move on. This is the advice Bucky constantly gave his audiences, and I feel that he qualifies as a sage elder worthy of consideration and respect. That's the path I have chosen, and I invite you to consider joining me on this exploration to create Bucky's vision of *"a world that works for everyone."*

People living today know Buckminster Fuller as a wise elder, but the wisdom of our elders is missing in the majority of "modern, developed" cultures. In 1995, after studying and teaching about Bucky's ideas for more than a decade, I realized that few elders were being recognized and admired as role models. Bucky Fuller was one of several such men and women who had served as role models and teachers for myself and many people, but I was concerned that my children did not have popular public archetypical elder leaders and mentors.

When I saw Bucky speak to thousands of people (mostly under the age of thirty) at the World Symposium on Humanity in 1979, he was eighty-three, and he exemplified my concept of a wise visionary elder. His vision rocked my reality, and I soon realized that he was not some pie-in-the-sky, optimistic intellectual.

Sitting alone on stage was an eighty-plus-year-old prophet who claimed he was simply *"an average healthy human being doing what he saw needed to be done on behalf of all of humanity."* He had, however, done his research and comprehensive thinking so that he could create solutions to problems he knew would confront society fifty to one hundred years in the future. We are the beneficiaries of those efforts. He also had the foresight to take his ideas one step farther by designing and building inventions that solved many of those yet unnoticed or undefined problems.

Bucky labeled himself a *"comprehensive anticipatory design scientist"* who offered us the opportunity to choose utopia at this critical juncture in our evolution, which he described in his book *Utopia or Oblivion*. He was the wise elder that I had wanted to study with, but he died before my wish came to fruition.

Fortunately, Bucky was constantly reminding us that he was not *"yesterday's breakfast."* In other words, he was not his physical body, and the fact that the body we called R. Buckminster Fuller no longer exists in physical reality on Spaceship Earth does not mean that he is gone.

We still have his massive archive of wisdom in many forms including writings, recordings, inventions, and the numerous people who have studied his work and continue to champion his vision. This book contains insightful commentaries from some of the people who carry on his legacy. Many of us consider ourselves to be some form of comprehensive anticipatory designers / scientists / artists / businesspeople / teachers / parents / writers / etc. because we know that the world needs visionaries in all areas of life if "we the people" are to survive and thrive as a species.

I tend to label myself a "comprehensive, anticipatory designer in training" with an aspiration to contribute as much of what I have learned to as many people as possible. This book is one physical artifact in my modest campaign to share the bounty of Bucky's wisdom and teachings with others. It also contains interpretations of Bucky's words as they have continued to flow from the "pattern integrity" that has been labeled with several names including Buckminster Fuller.

Bucky taught that all wisdom and knowledge is constantly available to every person if we tune into it. Sometime between my first contact with Bucky Fuller in physical form and his death in 1983, I began adjusting my "receiver / tuning device"

to more clearly pick up and record the pattern integrity frequencies being broadcast by and through Bucky. This put me in the odd position of being someone who did not agree with the commonly accepted view that an extremely important loss happened on July 1, 1983 when the person known as Buckminster Fuller consciously left his body for the last time.

On that day, Bucky experienced a significant transformation, but he was far from gone. He has not left us, and he continues to influence much of our reality today. After July 1, his staff and family packed up the massive physical artifacts he had accumulated. They believed that the experiment he had initiated in 1927—to determine and document what one individual could achieve that could not be accomplished by any institution no matter how large or powerful—was complete.

I felt that the next phase of Bucky's experiment had just begun, but few people were interested. Remembering Bucky's admonition that *"you can make money or you can make sense"* (covered in Chapter 3), I decided to follow in his footsteps, and I formulated the intention of making sense in the exploration and sharing of Bucky Fuller. That intention resulted in the writing of *Buckminster Fuller's Universe*, which was first published in 1988; the writing of *A Fuller View* and the future volumes currently being produced; and my sharing Bucky's wisdom in live and recorded presentations.

One factor that influenced me in this somewhat intimidating decision was Bucky's teaching that humans are part of Nature and Nature always supports what is supposed to be. Following that logic, I decided that if this work was destined to be shared with others, it would be supported. With that insight as my foundation, in 1984, I embarked on another phase of Bucky's grand experiment, and I began an extensive examination of the data he had amassed. I also continued to refine my ability to tune into the frequency of the pattern integrity that had inhabited a physical body between 1895 and 1983 and been known as Bucky Fuller.

For me, an experiment is not over once the data has been compiled. That data needs to be examined and evaluated so that it can be purposefully applied in the future. In the instance of Bucky's "56 Year Experiment," I have found an enormous number of simple solutions to all types of problems. His experiment covered everything from dealing with daily life issues to solving all humankind's physical problems while attempting to be as conscious and as cooperative as possible. All this and more is available through the wisdom of this wise elder who did everything that he could to support abundance and success for all.

A *Fuller View* was written to open new doors and awarenesses for anyone who reads even a small section. It will support us in remembering that we have come here at this moment in time to be transformational agents creating sustainable solutions as we steward our beloved Mother Earth on behalf of all beings.

I took on this task because lives are transformed in Bucky's presence because of the questions he asked and the intention he kept at the forefront of all his work. He formalized the great cosmic question that fueled his vision one last time just months before his death when he wrote,

> *"Human integrity is the uncompromising courage of self determining whether or not to take initiatives, support or cooperate with others in accord with 'all the truth and nothing but the truth' as it is conceived by the divine mind always available in each individual.*
>
> *Whether humanity is to continue and comprehensively prosper on Spaceship Earth depends entirely on the integrity of the human individuals and not on the political and economic systems.*

*The cosmic question has been asked — ARE HUMANS
A WORTHWHILE TO UNIVERSE INVENTION?"*

*February 14, 1983 — Penang, Malaysia.
St Valentine's Day and the Chinese New Year*

On that auspicious Valentine's Day at the beginning of another New Year, he reminded us where we are in Universe, and it is my aspiration that this quote is the foundation of everything that follows. Here Bucky repeats the cosmic question I posted on my bathroom mirror, *"Are humans a worthwhile to Universe invention?"* More important to each of us is the inquiry, "Am I a worthwhile to Universe invention?"

Forever the questioning optimist, Bucky constantly inquires about the things we all need to ask ourselves as often as possible. Am I living my life in harmony

–with Universe

–with the true abundance now available to all people

–with *"the truth and nothing but the truth as is conceived by the divine mind always available in each individual?"*

And how do you answer the other important questions:

• Are you doing what you see that needs to be done and is not being attended to?

• Are you a global citizen cooperating with others and Nature to further create a sustainable environment that supports all life?

• Are you willing to have an abundant life free of violent conflict and weaponry?

• And are you committed to leaving a legacy of a more sustainable, cooperative planet for future generations?

My response to these and the issues advanced throughout *A Fuller View* by the many global visionaries who so generously contributed to this book is that I am doing my very best. And that's what we can each do—our very best as mindfully and compassionately as possible. These are not easy questions, but we have chosen to live at a time when they are at the forefront of our survival. My aspiration is that the following pages will support you and all readers to answer these questions with a resounding "*YES!*"

"Friend to all the Universe" is one phrase used to describe Bucky in song and verse. May each of us be a friend to all the Universe so that we experience Bucky's vision of "*a world that works for everyone*" in our lifetime.

L. Steven Sieden
Seattle, Washington, Earth
July 1, 2011— 28 years after Dr. R. Buckminster Fuller's death on July 1, 1983

# BRIEF BIOGRAPHY OF DR. R. BUCKMINSTER FULLER

 RICHARD BUCKMINSTER FULLER WAS BORN ON JULY 12, 1895, in Milton, Massachusetts, and died on July 1, 1983, in Los Angeles. The eighty-eight years of his life provide us with a rigorously documented example of what one "average healthy man" can achieve when he or she has a clear intention, an open mind, a broad perspective, and the integrity to follow his or her heart.

Fuller's distinguished New England lineage included his aunt, the well-known transcendental feminist Margaret Fuller. The family's history in America dated back to Fuller's great-great-great-great-grandfather, British Navy Lt. Thomas Fuller, who traveled to the American colonies in 1630. Like previous generations of Fuller men, Bucky's father, Richard Buckminster Fuller Sr., graduated from Harvard, but he was the first Fuller male in eight generations who did not become a minister or lawyer. So, it would seem that Bucky's future path was established well before he was born.

As a Fuller man, he was expected to graduate from Harvard and become a minister or a lawyer. Bucky, however, had other ideas, and nothing could be farther from his life path than those two professions. Over the course of his life he often criticized organized religion (not spirituality) and the competitive corporate system, which he felt was dominated by greedy lawyers.

Bucky was a headstrong person who knew what he wanted and was willing to take enormous risks to achieve his goals, which were comprehensive and included the welfare and success of all life on the planet he named Spaceship Earth. Still, even in the last years of his life, he remained a humble *"average man."* Despite the fact that he was a highly respected and honored global elder who had been presented with hundreds of awards and dozens of honorary degrees, he insisted that everyone call him "Bucky."

From the early 1930s through his death in 1983, Bucky was an increasingly well-known global visionary with a twofold mission.

1) To demonstrate and document what one individual could achieve that could not be accomplished by any institution or organization no matter how affluent or powerful.

(2) To advocate and work for the success of all life on the planet he named Spaceship Earth using his Comprehensive, Anticipatory Design Science.

Long before the term ecology became a household word, he asserted that humankind was on the verge of an enormous shift in which resources had to be consciously recycled in order to survive and thrive. He also collected the data to prove that when we recycled resources we were doing more with less and, thus, would be able to eventually do so much more with so much less that we could support everyone on Earth.

Bucky's contribution to humankind has yet to be fully appreciated. The fact that he consciously documented every aspect of his life provides us with an enormous amount of practical information. By examining his 56 Year Experiment on behalf of all humankind, we can begin to more clearly uncover what works and what doesn't in living a life that makes a conscious positive difference.

△ △ △

This biography separates the multitude of Bucky's experiences into four distinct periods. The first of those periods was from his birth in 1895 through 1927. That was a time of experimentation and learning. He extended himself and his environment outward in search of limits; often learning from the painful lessons most people would categorize as failure.

It was also a time of youthful exuberance and disappointment. Fuller suffered the loss of his father, graduated from Milton Academy, twice enrolled at and was expelled from Harvard University, and worked as an apprentice millwright in Canada.

Then he served as an officer in the Navy during World War I, married Anne Hewlett, and experienced the birth and untimely death of their first daughter. As a businessman, he held managerial positions in several diverse corporations and organized his own construction company. He used that company, Stockade Systems, to attempt establishing and propagating a radical new form of construction that failed financially after a few years. As a result of that endeavor, by 1927 he had lost all his money as well as the investments of his friends and family.

1927 marked the major shift in Bucky's life and path as well as the beginning of the second period of his life, which lasted until 1947, when he invented the geodesic dome. With the loss of his construction company and the birth of his second daughter, Allegra, Bucky found himself stranded with a young family in 1927 Chicago. He had no money, no job no formal education beyond high school, a reputation as an unsuccessful businessman, and no prospects for the future.

Extremely dejected, he seriously considered drowning himself in Lake Michigan. It was then that Bucky had the famous mystical experience that transformed his life. He realized that he did not belong to himself and, consequently, did not have the right to end his own life.

In that cosmic flash, Bucky suddenly understood that he (like every human being) belonged to Universe, and he committed himself to an experiment that provided the foundation and context for his every action and decision during the next fifty-six years. He decided to embark upon a lifelong experiment to determine and document what one average, healthy individual with no college degree and no money could accomplish on behalf of all humankind that could not be achieved by any nation, business, organization, or institution, no matter how wealthy or powerful.

With no apparent means of support for his family much less his experiment, Bucky resolved to use the only person available for observation. Thus, R. Buckminster Fuller adopted the alias "Guinea Pig B," one person under the constant scrutiny of himself.

During the twenty years from 1927 through 1947 a more mature Fuller devoted a great deal of his time to a formidable search for Nature's coordinating system. The discoveries he made in that investigation eventually became his radical Synergetic Mathematics, a mathematical system based on what he observed in Nature rather than man-made ideas and concepts. That system also became the foundation for Bucky's most famous invention, the geodesic dome, and other of his insights and creations.

Following his commitment to work on behalf of all humankind and to never again work for a living, the initial problem Bucky took on was the issue of housing the expanding population of Spaceship Earth. He worked sporadically on the mass-producible Dymaxion House from 1927 until Beech Aircraft and the United States government joined him to produce a prototype in 1944.

During the 1930s Bucky held positions at Phelps Dodge and *Fortune* magazine, and these provided him with an opportunity to study Earth's resources and Nature's efficient operating strategy of doing "more with less" resources. Once he realized the significant benefit of Nature's way of doing everything, he adopted it as a primary aspect of his work and life. Years later, he would bring the phenomenon of doing more with less into the popular culture as "synergy," a term he singlehandedly moved from the obscurity of the chemistry lab into the light of public awareness.

In 1947, during one of his many stints as a visiting college professor, he combined his mathematical skills with his knowledge of construction and Nature's coordinating system to produce his most famous invention—the geodesic dome. The creation of the geodesic dome also ushered in the third significant period of his life, which spanned from 1947 through 1976 as he continued his explorations while attaining celebrity status for his work with geodesic domes.

In the 1955 at the age of sixty, Fuller could have retired on his licensing fees from the geodesic dome, but he had no interest in gaining great wealth or slowing down. Instead, he expanded his effort to create success for all humankind.

He personally designed and supervised the construction of most of the significant geodesic domes built during that period. He also expanded his unique "thinking out loud lecturing" around the world. In response to a constant flow of invitations, he was making at least 130 such appearances every year. In fact, he continued that hectic speaking schedule until his last presentation, an all-day session focused on integrity to a sold out auditorium in Huntington Beach, California, just one week prior to his death.

Bucky's campaign on behalf of the success of all humans and life on Spaceship Earth was the focus of the last phase of his life from 1976 until his death on July 1, 1983. During that period, he was continually traveling, making presentations, writing and sharing as much of what he had learned as possible. It was a last ditch effort to make certain that his life was complete and that the had given everything possible in support of his mission to create *"a world that works for everyone."*

Bucky wrote and published a detailed explanation of all that he had learned and found to be true regarding the human experience in his final major book, *Critical Path*. It outlines the course humankind must follow if we are to survive and thrive, and it explains that we have entered into a new era of sufficiency, you *and* me, cooperation and peace in which previous solutions and behaviors such as war, competition, politics, and corporate domination are obsolete.

This is the wisdom and the challenge that R. Buckminster Fuller left for us. His insights are as much a shining gem of hope and possibility as they were when he traveled around Spaceship Earth sharing his positive perspective that we can succeed as a species and be good stewards of our planet if we cooperate and shift our resources from weaponry to "livingry."

Bucky remained true to his mission for 56 years. During that period, he saved and archived every possible aspect of his life, creating his Chronofile and making his life the most documented of any "ordinary, average" person in the history of humankind.

Although his personal experiment has yet to be fully examined, the success of Bucky's life is indisputable. After discovering many of the natural underlying principles that govern all Universe, Bucky applied them to every aspect of his work where he:

- Was granted twenty-five U.S. patents.
- Wrote twenty-eight published books and thousands of articles.
- Received forty-seven honorary doctorates.
- Was presented with hundreds of major awards.
- Circled the globe fifty-seven times working on projects and lecturing.

- Presented an average of one hundred "thinking out loud" sessions per year (often labeled lectures, they would range from two to six or more hours in length), even when he was in his eighties.

Most important was his documentation and demonstration of the importance of the "little individual" in the grand scheme of human evolution. Living as a global citizen, Bucky was able to teach by example—showing us with his accomplishments and seeming failures that each of us possesses tremendous gifts that we can contribute to others and help create *a world that works for everyone.*" He also proved that a person could have a satisfying, enjoyable life while making his or her unique contribution.

Dr. R. Buckminster Fuller died on July 1, 1983, at Good Samaritan Hospital in Los Angeles, where he was faithfully watching over his beloved wife, Anne, who was in a coma and not expected to regain consciousness. Sitting at her bedside, hand in hand with his wife of sixty-six years, he felt something and exclaimed, "She squeezed my hand!"

Moments later, Bucky experienced massive heart failure and died. Anne never regained consciousness and died thirty-six hours later. They were buried together in Milton, Massachusetts under a tombstone that reads "Call Me Trimtab."

The conscious nature of his life and death became more obvious when a neat stack of papers was found on his desk. The note on top of those papers asked the finder to please make sure that this material was published, as it was the final thing Bucky had to say.

Although that material was published years later as the book *Cosmography*, it does not seem to be the last thing Bucky has to say. For many of us, he continues to speak and remind us again and again that humankind has entered its *final examination*" and that we can survive and thrive if we begin cooperating with one another and working for the betterment of all life on Earth.

# Highlights of "An Average Man's" Life

## Chronology of the Life of R. Buckminster Fuller

1895—Richard Buckminster Fuller is born in Milton, Massachusetts.

1899—Enters kindergarten where he builds the first octet truss using dried peas and toothpicks.
Is diagnosed as nearsighted, gets his first glasses and sees objects clearly for the first time in his life.

1904—Enters Milton Academy, Lower School.

1910—Bucky's father, Richard Buckminster Fuller Sr. dies.

1913—Graduates from Milton Academy and enters Harvard University (Class of 1917).

1914—Expelled from Harvard and sent to be an apprentice millwright in remote Canadian mill.

1915—Reinstated at Harvard and expelled for a second time. Takes a job with Armour & Co in New York.

1917—Enlists in Navy Reserve.
Marries Anne Hewlett.

1918—Promoted to lieutenant and assigned as aide to admiral commanding all transports in the Atlantic during World War I. First child, Alexandra, born.

1919—Appointed Communications Officer on the *USS George Washington* and supervises President Woodrow Wilson having the first ever transatlantic radiotelephone conversation (from France to the United States).

1922—Alexandra dies in his arms just prior to her fourth birthday. Bucky feels responsible for her not having better housing and he begins his lifelong quest to provide excellent shelter for all people.

Begins working as an entrepreneur, founding Stockade Building Corporation, manufacturing buildings with a revolutionary new technology.

1926—Is fired as president of Stockade Systems when the company is consistently unprofitable because Bucky has chosen to build good buildings rather than make a profit.

1927—Second child, Allegra, born.

Considers himself a failure and contemplates suicide. Has mystical experience in which he is told that he does not have the right to kill himself and that he will only speak the truth from then on. Dedicates himself to the service of all humanity. Writes and privately publishes first book, *4D Timelock*.

1929—Displays model of 4D round house at Marshall Field Department Store. Coins and copyrights the word *"Dymaxion"* to describe house and other inventions.

1933—Establishes Dymaxion Corporation to successfully design and build first prototype Dymaxion Vehicle.

1935—Prototype Dymaxion Vehicles #2 and #3 are completed and displayed at the Chicago World's Fair.
Writes *Nine Chains to the Moon*.

1936—Meets with Albert Einstein who is amazed that Fuller could conceive of a practical application for Einstein's theories.

1938—*Nine Chains to the Moon* published.

Joins *Fortune* magazine as science and technology consultant.

1940—Works on development of Dymaxion Deployment Units at Butler Manufacturing in Kansas City.

1941—Quits drinking and smoking as an anniversary gift for Anne and to further his mission without his behavior as a hindrance.

1942—Joins US Board of Economic Warfare as Director of Mechanical Engineering.

1943—Full-color, punch-out rendition of *Dymaxion Sky-Ocean World Map* is published in *Life* magazine resulting in the largest printing of the magazine.

1944—Begins design and production of prototype Dymaxion House in conjunction with Beech Aircraft of Wichita, Kansas.

1946—Is awarded the first cartographic project patent since 1900 for Dymaxion Map.

1947—Invents Geodesic Dome.
First teaches at Black Mountain College.

1948—Teaches at MIT.

1949—Begins extensive travels responding to speaking invitations worldwide.

1952—Begins work on Ford Motor Dome in Detroit.
Constructs first Geoscope with students at Cornell University.
Receives Award of Merit, American Institute of Architects.

1954—Receives patent for Geodesic Dome.
Receives first of his eventual forty-seven honorary degrees, Doctor of Design from North Carolina State University.
Asked to design a geodesic dome to cover Dodger Stadium.

1955—First Distant Early Warning (DEW) Line Radomes are installed in Northern Canada.

1956—Becomes visiting lecturer at Southern Illinois University.

1957—Union Tank Car Dome in Louisiana and Kaiser Hawaii Symphony Dome are erected.

1958—Makes the first of many annual circuits traveling around the world speaking (primarily at universities).

1959—Appointed professor and awarded honorary doctor of arts degree at SIU, Carbondale IL. Sets up global headquarters at SIU.

He and Anne move into a Carbondale geodesic dome home.

1961—Granted patent for octet truss he first built in 1899 at the age of four.

1962—Appointed as Charles Eliot Norton Professor of Poetry, Harvard University.
Establishes "Inventory of World Resources, Human Trends and Needs" (World Game) at SIU.
Granted patent for tensegrity structure.

1963—Publishes the books *No More Secondhand God, Ideas and Integrities,* and *Education Automation.*
Appointed to NASA's Advanced Structures Research Team, which adopts his octet truss and geodesic dome as primary space structures.

1964—Subject of *Time* magazine cover story.
Publishes *Design Science Decade: World Inventory, Human Trends and Needs.*

1966—Completes design for USA Pavilion at '67 Montreal World's Fair.
Inaugurates World Game at SIU.

1967—Featured in *Saturday Review* cover story.
Montreal Expo Dome draws record attendance of 5.3 million people in six months.
Elected to honorary membership in Phi Beta Kappa (Harvard) on the occasion of the fifty-year reunion of the Class of 1917 (from which he was expelled in his first year).
Granted patent for star tensegrity.

1968—Elected to National Academy of Design and World Academy of Arts & Sciences.
Appointed Distinguished University Professor at SIU

1969—Leads first public World Game workshop.
Delivers Jawaharlal Nehru Memorial Lecture.
Publishes *Operating Manual for Spaceship Earth* and *Utopia or Oblivion.*

1970—Publishes *I Seem to Be a Verb*.
Installed as Master Architect for Life by Alpha Ro Chi Architectural Fraternity.

1971—Presents proposal for "Old Man River City" in East St. Louis, and it is accepted.
NBC broadcasts documentary *Buckminster Fuller on Spaceship Earth*.

1972—Appointed "World Fellow in Residence" by a consortium of Philadelphia institutions.
Publishes *Intuition* and *Buckminster Fuller to the Children of Earth*.
Delivers over one hundred twenty "thinking out loud" lectures around the world.
Is featured as interview in *Playboy* magazine.

1973—Establishes publication and research office in Philadelphia.
Granted patents for floating breakwater and tensegrity dome.
Moves to Philadelphia.

1974—Delivers over one hundred fifty "thinking out loud" lectures around the world.
Granted a New York State architect's license.

1975—Publishes *Synergetics, The Geometry of Thinking*.
Is elected International President of the World Society for Ekistics.
Granted patent for non-symmetrical tension-integrity structures.
Inducted as a fellow in the American Institute of Architects.

1976—Participates in drafting "Declaration of Principles and Rights for American Children."
Publishes *And It Came To Pass—Not To Stay*.
Completes work on world's first tetrahedronal book, *Tetrascroll*, which is published in limited edition.

1977—Designs and develops two new prototype geodesic domes, "Pinecone Dome" and "Fly's Eye Dome."

Travels on lecture of Far East sponsored by US government.

1978—Appears in ad for Honda Civic.

Becomes Senior Partner in NY architectural firm Fuller & Sadao.

1979—Publishes *Synergetics 2*, expanding on *Synergetics* and *Buckminster Fuller on Education*.

Becomes Chairman of the Board of R. Buckminster Fuller, Sadao & Zung Architects of Ohio.

Becomes Senior Partner of Buckminster Fuller Associates, London, England.

1980—Granted patent for tensegrity truss.

Delivers over ninety "thinking out loud" lectures around the world.

Moves to Pacific Palisades, CA, near his daughter Allegra and her family.

1981—Publishes *Critical Path*.

Inducted into the "Housing Hall of Fame."

1982—Publishes three books: *Grunch of Giants, Inventions,* and *Humans in Universe*.

Designs and supervises Dymaxion "Big Map" the size of a basketball court and displays it to members of the US Congress.

Delivers over seventy "thinking out loud" lectures around the world.

Inducted into the Engineering and Science Hall of Fame.

Granted patent for hanging storage shelf unit.

1983—Honored in daylong "Integrity Day" presentations in several US cities.

Dies July 1 in Los Angeles, CA.

# ROADMAP TO UNIVERSE

## LOGISTICS & STRUCTURE OF THIS BOOK

 *"We have to be comprehensive in our thinking, to be infinitely tolerant, to understand rates of change, to understand the inertia, to understand the fear. We must avoid doing things that will excite the fear and do everything that will eliminate it. I don't mean then to eliminate challenge-- for challenge we must have. We will continue to have challenge because Universe is continually changing, and we are going to continue to be confronted with the new."* – Buckminster Fuller

THIS SOMEWHAT SUCCINCT AND EASY-TO-UNDERSTAND (FOR Buckminster Fuller) quote expresses the context of this book. *A Fuller View* is designed to support us all (including me, the author) in eliminating fear and handling the challenges that face us both individually and as a global society. The quotes selected in this book are both profound and easy to digest quickly. Although they are some of Bucky's best, they in no way come close to the complex intensity that pervades the majority of his writing and speaking. The following is a more typical Bucky sentence / paragraph selected from Chapter 2 of Fuller's diminutive book *Grunch of Giants:*

*"In contradistinction to brains, which are constructed of physical matter, the weightless, matterless, metaphysical mind has the unique capability, from time to time, to discover eternal interrelationships existing invisibly between*

*special-case experiences, which interrelationships cannot be discovered by any or all of the brain's physical sense systems--for instance, the mathematical law governing gravity's invisibly cohering not only the Sun and its multi-millions-of-miles-apart planets as discovered by the weightless invisible conceptual thought-relaying from the mind of Kepler to Galileo to Newton and as also cohering the never-anywhere-intertouching parts of local Universe systems of galaxies, and electrons remote from their nuclei."*

If you spend as much time as I have with Bucky and material such as this sentence, it will eventually morph into something very clear and extremely relevant. In today's world, however, most of us don't have the time, much less the interest, to interpret massive sentences like the one above, which describes the distinction between brain and mind while providing a historic scientific example of human mind in action. Accordingly, the quotes chosen will most likely not cause your head to spin and will provide insights into the mind and experiences of a great genius.

For most people, the pages that follow will offer new ideas, perspectives and challenges. I have endeavored to make your journey through the comprehensive, anticipatory design science Universe of Buckminster Fuller as valuable and efficient as possible with this *Roadmap to Universe* explanation of the logistics and quirks presented in these pages.

*A Fuller View* is not a normal book, and I did my best to make your journey within it as smooth and useful as possible. Because of the comprehensive nature of Bucky's life and vision, the creation of a 270-page book that includes all the key points of his philosophy is daunting if not impossible. Reading it may be similarly intimidating, so here are some tips for your journey.

1. *A Fuller View* was not written to be read cover to cover from beginning to end. It was designed as an inspirational resource that could be opened anywhere and understood. Or opening to a random page might pose ideas to contemplate. Once I began adding the guest commentaries, *A Fuller View* became even more complex and less linear. I did my best to create a flow from quote to quote and chapter to chapter, but I chose the best content even if it did not create the best flow.

2. Although I have endeavored to organize the quotes and commentaries in a standard chapter arrangement, most of them could have been placed in several of the chapter categories. The arrangement is therefore simply another tool to support readers in navigating their way through and around Buckminster Fuller's wisdom and reality.

3. Guest commentators were asked to pick their favorite Bucky Fuller quote. That resulted in adding some amazing insights that would otherwise never have been included. It also created a more interesting melding of ideas, some of which had not crossed my mind in a long time.

4. Many of the ideas and insights appear several times throughout this book. They may be written in a different way or you may find the exact same words. I have not counted the number of times "a world that works for everyone" or "Call me Trimtab" appear, but I know that these and other critical elements of Bucky's grand vision appear numerous times.

Rather than ignore them as "something I already know about," when they come up a second or third or tenth time for you, I suggest that you consider them a reminder of things that we "know we don't know." Or you could think of them as gifts from Universe returning to support you in waking up.

They appear because of their significance. Trimtab, more with less, ephemeralization, a you and me society, a world that works for everyone, shifting from weaponry to livingry are keys to Bucky's work and wisdom as well as to the survival and success of humankind. Thus, it's appropriate that these ideas appear several times.

5. In creating some type of effective efficient structure for the wisdom of Buckminster Fuller, which on the surface appears to be vast and unstructured, I found it best to "go with the flow" and realize that "change is inevitable." Still, I did follow Bucky's sense of respectful New England politeness, so Guest Commentaries always appear before my writing on a quote. When more than one Guest wrote on a particular Commentary, the order of the Guests is random. I did put the father John Robbins before his son Ocean as the two commentaries seemed to flow in that way, but the order of others was randomly chosen--if there is such a thing as random in Universe.

6. All the Guest Commentator's essays are prefaced with a byline and highlighted by a gray stripe down the side. To help you distinguish where they stop and my writing starts, my writing always begins with △.

7. Bucky's "56 Year Experiment" is referenced numerous times throughout this book because it is so important to his work and wisdom. In 1927, Bucky contemplated suicide but was "instructed" by a voice (inner wisdom, epiphany, guide or whatever label you choose) that he did not have the right to kill himself He then did what so many master teachers do by going into a period of voluntary silence and considering what to do next.

That "next step" for him was an experiment using himself as "Guinea Pig B." Bucky describes it on page 124 of Critical Path when he writes:

*In 1927, at the age of thirty-two, finding myself a 'throw-away' in the business world, I sought to use myself as my scientific 'guinea pig' (my most objectively considered research 'subject') in a lifelong experiment designed to discover what--if anything--a healthy young male human of average size, experience, and capability with an economically dependent wife and newborn child, strong without capital or any kind of wealth, cash savings, account monies, credit, or university degree, could effectively do that could not be done by great nations or great private enterprise to lastingly improve the physical protection and support of all human lives, at the same time removing undesirable restraints and improving individual initiatives of any and all humans aboard our planet Earth."*

My less detailed description of the "56 Year Experiment" is that Bucky made a lifelong commitment to determine and document what one average individual could achieve that could not be accomplished by any government, corporation, religion or other institution, no matter how rich or powerful. For me the critical aspect of his commitment was the documentation. Bucky left us an enormous archive—a genuine legacy of one man's quest on behalf of all humankind.

In my experience, that experiment has been largely ignored for the past twenty plus years. As with any experiment, it needs to be examined to see what worked and what did not so that we can employ the insights in our lives and as a society. That is a primary aspect of *A Fuller View*. It's one small step toward examining the "56 Year Experiment" and providing another doorway into applying its wisdom.

8. The previous quotation and the one about brain / mind on page xxxvi demonstrate a primary motivation for the creation of this book as well as my earlier Bucky book, *Buckminster Fuller's Universe* (Basic Book 2000). Sentences that are paragraphs represent Bucky's style of writing and speaking. Thus, when people ask me, "What would you recommend I read to learn more about Bucky?" my reply is always the same.

"You don't read Buckminster Fuller, you study him, and I recommend that people study his writing in a group." This is the material that makes for a deep book club. I spent every day for three months on a California beach doing just that, and I now realize that my experience would have been much more productive with the support of others. I did have the spiritual support of Bucky himself during those months in Laguna Beach, but it was not an easy process.

As the saying goes, "How do you eat an elephant? One bite at a time." A *Fuller View* will help you bite off and digest small pieces of the "elephant" known as Dr. R. Buckminster Fuller's wisdom, experience and insights. I am hopeful that it will support you and all readers to enjoy the treats rather than choking on the enormity of the gifts Bucky provides.

9. All words spoken or written by Bucky (except the 55 quotes that make up the body of the book) are *italicized* for clarity. Hopefully, this unusual formatting will not be confusing as I also had to use standard formatting and italicize things like book titles and foreign words. Many of the Guest Commentators used italics for emphasis, and I did my best to change them to bold or some other form for clarity. That, however, was not always possible.

10. "Universe" is almost always capitalized and without the article "the" because that is the format Bucky used, and I follow in his path. His rationale is simple, and I agree. There is but a single Universe that includes everyone and everything. By capitalizing the word "Universe," we elevate Universe to the level of humans, planets and corporations--all of which have capitalized names. By eliminating the article "the," we no longer relegate Universe to being one of many universes in existence.

11. "Earth" is also capitalized throughout this book. The following statement that Bucky wrote as a preface for his Introduction to the 1977 book *American Space Photography* explains why.

*"As I write the introduction to this truly fascinating book, I find its author and its publishers spelling our planet's*

name with a small 'e,' even though its 7,926 mile diameter is larger than the mile diameters of Venus's 7,620, Mars' 4,220, Pluto's 3,600 or Mercury's 3,010. The first letter of all the other planets' names we honor with capital letters. I am confident that those who spell Earth with a small 'e' are not yet 'seeing ourselves as others can see us' from elsewhere than aboard our cosmically miniscule planet."

12. One aspect of our language that continues to confound me is what I label the "he / she conundrum." When writing about generalized people, an author has to choose either "he," "she" or the now more popular but plural "they." None of these options appeal to me, so throughout this book you will find "he" or "she" used randomly when necessary.

13. "Guest Commentators" were initially not part of this book. Now, they represent nearly half of the content, and I am extremely grateful for the generous contributions these people have made to the world and this book in particular.

The idea for Guest Commentators was a "cosmic fish" that I feel was sent floating through my mind by Bucky several times before I noticed it and several more times before I took action. I now feel that he wanted his friends to once more come together in creating a synergistic whole much greater than its parts and supporting an eternally regenerative Universe and Spaceship Earth.

When I approached the publisher Michael Wiese with the idea of Guest Commentators, he was thrilled and wrote it into our formal agreement. At the time, I reluctantly agreed in writing to find at least twelve Guest Commentators. Then, the process began and the floodgates of support opened. People I invited were extremely supportive and began writing back, "Have you asked my good friend _____. She should really be part of this."

The result was 130 invitations made over the course of only a few months and the 42 Guest Commentators who generously responded with 32,000 words of wisdom for us all. Some of the most rewarding gifts I have received since this process began have been these essays arriving in my inbox.

Reading them as they arrived inspired me to continue what has often felt like running a marathon, and I am grateful beyond words to these very generous people. I'm also grateful to everyone else who responded to the Guest Commentator Invitation because nobody said "no" or even "no thank you." Almost everyone who declined did so because they could not fit participating into their busy schedule. In other words, they said "not now," which left me thinking about Volume #2.

14. Selecting the quotes included in this volume from the tens of thousands possible was both challenging and flowing, especially once the Guest Commentators began sending in contributions. Then, it became a question of which quotes and my commentaries that had already written would be edited out. In the final analysis, 30,000 plus words and quotes were set aside for another day. In other words, I edited out half of my own writing. Again, Volume #2.

I also felt guided by Bucky's voice, hand and spirit throughout this process. As a few of the Guest Commentators have mentioned, it feels like Bucky is continuing to share his wisdom even through he no longer operates through the physical body we labeled Buckminster Fuller. It's just a matter of tuning into the frequency of that pattern integrity known as Bucky Fuller.

As Dr. Cherie Clark wrote in her essay,
"I started writing my doctoral dissertation, 12° Of Freedom: Synergetics and the 12 Steps to Recovery, on nights and weekends from 1996 – 2001... There were countless times that I'd think, 'I know Bucky said ... about this.' I could hear him saying it, and wanted to make sure I had the exact quote. I'd reach over to my pile of 'Bucky books,' pick up one and open it to the

exact quote I was looking for. At first, it startled me. Then, it became a game. Usually around 3 a.m. I would hear Bucky's distinctive voice, with a suggestion for how to more clearly respond to a question I was wrestling with earlier. He was right beside me the whole time I was writing, more than 13 to 18 years after he had departed this life, guiding, coaxing, teasing me to understand what he had said."

14. Each chapter ends with a "Conclusion" quote and commentary. These are by no means meant to be a summary of the chapter, but rather a "final thought" that may provide a sense of direction for the reader. The same is true of the initial quote, which is designed to set context for the balance of the chapter. It is not necessary to read either the initial or final pieces in order to receive the full benefit of any of the material in between.

15. Dr. R. Buckminster Fuller preferred to be called "Bucky." This was a choice he made at a very young age, and he continued on with that tradition even in the last years of his life when many of us felt moved to call him Dr. Fuller, Sir, Mr. Fuller or something that demonstrated our respect for him as a wise elder. Still, he always asked to be called Bucky, so that's the name I use most often in this volume.

16. This book is designed to be enjoyable, inspiring and educational. Please, don't take it too seriously. We've reached a time in our evolution when "we the people" can take charge, and every individual does make a difference. Still, it's also necessary to relax, enjoy and let go. Then, each of us can go with Nature's flow as Bucky did for most of his life.

CHAPTER 1

# VIEW

## UNIVERSAL
## PERSPECTIVE

## 1.1 "I look for what needs to be done and then try to work out how to do it best. After all, that's how Universe designs itself."

DOING WHAT NEEDED TO BE DONE WHILE ADHERING TO Nature's principles was the strategy that Buckminster Fuller used to support the creation of his vision of *"a world that works for everyone."* By employing trial and error experimentation and making lots of "mistakes" (i.e. learning experiences), he was able to achieve a great deal, and even today nearly thirty years after his death, his legacy continues to make a difference in the lives of millions of people.

If each of us would adopt even a tiny fraction of what Bucky learned and shared throughout the eighty-eight years of his lifetime on Spaceship Earth, then our world would be a much better place to live. If we can simply begin to notice how Universe / Nature / God / Higher Power / Great Spirit / Allah designs itself and follow just a small portion of that design, we will markedly transform ourselves as individuals and as a global society.

This is the primary cause that Bucky championed throughout his "56-Year Experiment" in which he himself was "Guinea Pig B," the subject of the experiment to discover and document what difference the *"average little man"* could make on behalf of the most people. He used the classic trial-and-error method of discovery so that we no longer have to go down that ineffective path in doing our part to make a difference on Spaceship Earth. He allowed himself to be subjected to some severe ridicule in order to discover and display how Universe designs itself and what needs to be done to create a sustainable environment for future generations and ourselves.

He also planned and completed his initiatives so that he could do the most with the least amount of resources using the natural principle he labeled "trimtab" to be as effective and efficient as possible. Thus, everything he did appeared nearly effortless, beautiful, and in harmony with all life. The key to his success was discovering how Universe/Nature responded to an idea or action, and then he would follow Nature's template rather than follow the path put forth by most "leaders."

Today, we are blessed with having his ideas and experiences as a guide for our actions. We don't have to reinvent the wheel. Following in his footsteps and deciding from our personal experience what works, each of us can do what needs to be done in the most effective, efficient way possible. We can function with complete integrity knowing that Universe always supports what needs to be done so that if our actions and projects are in harmony with the sustainability of all life we will succeed just as Bucky was able to do.

## 1.2 "Start with Universe."

Throughout history, human civilizations have been guided by interpretations of the cosmic order. Our ancestors observed patterns in nature that profoundly influenced their beliefs and behaviors, enabling them to anticipate and synchronize with the cycles of life. By paying close attention to the world around them, countless generations developed reciprocal relationships with environments that enabled them to survive and thrive.

Today, modern technologies enable us to manipulate our surroundings in extraordinary ways. Yet they also isolate us, encouraging us to take the life-sustaining systems of our home planet for granted as endlessly exploitable resources and economic externalities. As specialized sciences increasingly seek to reduce existence to its component parts, the universe has seemingly been diminished to little more than physical properties, isolated interactions, and mathematical laws.

This materialist cosmology has effectively separated facts from value, imparting the overwhelming sense that Earth is a mediocre pale blue dot aimlessly wandering around within the infinite void of space. Though the interconnected challenges facing humanity are growing ever more complex and urgent, there seems to be little guidance or meaning to be found by paying attention to this larger context.

Referring to humanity as "local-Universe information gatherers and problem-solvers," he strove to demonstrate how we are capable of comprehending the "relationship of eternal principles" and applying them *"in support of the integrity of eternally regenerative Universe."* Today this approach is

4    A FULLER VIEW        SIEDEN

recognizable within the field of biomimicry, though he set his sights even higher by exploring the possibilities of what could be called cosmomimicry.

Buckminster Fuller challenged this limited perspective over fifty years ago, asserting that it is both incomplete and obsolete. He insisted that the sense of separation from nature is a dangerous illusion resulting from reductionism and overspecialization and that humanity's evolutionary success is dependent on our willingness to learn from the emergent behaviors of whole systems. This led him to question how we envision the context of our existence, re-imagining a big picture in which our species is situated within the full continuum of creation. In *Operating Manual for Spaceship Earth* he asks,

*"Can we think of, and state adequately and incisively, what we mean by universe? For universe is, inferentially, the biggest system. If we could start with universe, we would automatically avoid leaving out any strategically critical variables. We find no record as yet of man having successfully defined the universe—scientifically and comprehensively—to include the nonsimultaneous and only partially overlapping, micro-macro, always and everywhere transforming, physical and metaphysical, omni-complementary but nonidentical events."*

Never one to shy away from a daunting task, Fuller redefined "Universe" (eventually differentiating it through capitalization and dropping the definite article "the") to include both the specialized insights of science and our metaphysical capacities and experiences. Yet he insisted that Universe is far more than simply mind plus matter, contending that the whole is always more than the sum of its reduced parts. He summarized this perspective with the pithy generalization U=MP, proposing that Universe is the synergistic result of the metaphysical multiplied by the physical (*Synergetics* 162.00).

This was far more than an intellectual exercise, as he sought to apply "macro-inclusive" and "micro-incisive" insights to the design of human-scale physical artifacts through what he called *"Comprehensive Anticipatory Design Science."* By spending much of his life starting with consideration of the biggest system, anticipating future trends and needs, and combining the aesthetics and intuition of design with the empirical and intellectual rigor of science, he took it upon himself to attempt to solve some of the greatest challenges he predicted would soon be facing humanity.

In the twenty-first century, this synergistic, systems-oriented approach is more critical than ever. At the Buckminster Fuller Institute, we are celebrating his legacy by connecting a global network of design science practitioners actively applying these principles in their own work. We are seeking out and cultivating integrated strategies designed to address Fuller's challenge to *"make the world work for 100% of humanity, in the shortest possible time, through spontaneous cooperation without ecological offense or disadvantage of anyone."*

Instead of defining a particular problem to be solved, we encourage participants to explore how the behavior of whole systems can inform the design of approaches that address multiple interconnected issues simultaneously. We never know what to expect, but we continue to be amazed and delighted at the extraordinary capacity humans seem to possess for applying the principles of nature to improve our world. Like Fuller, we anticipate that by paying ever-closer attention to Universe, our collective journey just might have a few more surprises in store:

*"I didn't set out to design a house that hung from a pole, or to manufacture a new type of automobile, invent a new system of map projection, develop geodesic domes, or Energetic-Synergetic geometry. I started with the Universe—as an organization of energy systems of which all our experiences and possible experiences are only local instances. I could have ended up with a pair of flying slippers."*

DAVID MCCONVILLE is creative director of the Worldviews Network (www.worldviews. net), a collaboration of scientists, artists, and educators re-imagining the big picture of humanity's home in the cosmos. Using immersive environments and interactive scientific visualizations, they are facilitating community dialogues across the U.S. about how our collective actions are shaping humanity's future. David is also President of the Buckminster Fuller Institute (www.bfi.org) and co-founder of The Elumenati (www.elumenati.com), a design and engineering firm specializing in the creation of immersive learning environments.

⚠ THIS SEEMINGLY SIMPLE STATEMENT REPRESENTS THE essence of nearly every one of Buckminster Fuller's initiatives, and it is key to the success he was able to achieve. He always did his best to mindfully begin from a perspective of the whole or, in other words, with Universe (which is more clearly defined later in this book).

By working at the very onset of an initiative within a context of the whole of reality (both physical and metaphysical), Bucky was able to create projects and artifacts that were far more valuable and viable than what most other people were doing. And in that initial aspiration of starting with Universe, he eliminated the often-used motivation that so many of us begin with—making money or making a living for ourselves and our families.

This is not to say that there is anything wrong with that intention, but rather that when coming from this narrow context, a person often misses the opportunity to really make a difference and succeed well beyond what she initially imagined. That shift in perspective can be seen in Bucky's famous statement, "*You can make money or you can make sense, the two are mutually exclusive.*" (Explained later in this book.)

On the surface, "*start with Universe*" appears at odds with another often-quoted statement that is not attributed to Bucky—"start with gratitude." Do we begin with the whole of Universe or with gratitude?

This is not an "either/or" question, but rather a "both/and" situation. We need to consider the whole of reality at the onset of an initiative (or at every moment of our lives), and we need to do it with the often-cited "attitude of gratitude." This is important because when a person considers the whole of Universe he can't help but feel gratitude regardless of external circumstances.

How can an individual not be grateful when she recognizes the vast abundance of Universe and the precious nature of human life? How can a person be ungrateful for any situation when he realizes that every set of circumstances—no matter how seemingly difficult or exciting—provides yet another opportunity for growth and learning? And how can anyone not be grateful when she "gets" the grand glory of the whole of Universe that has been provided as our experiential playground?

These are just a few of the inquiries that Bucky employed, and those of us who respect and utilize his grand strategy of life also use them to make a difference with our lives while enjoying every moment. If we do our best to observe and be in harmony with the whole of Universe with gratitude, we may just find ourselves living in a Universe of love, peace, joy, and freedom.

## 1.3 "Integrity is the essence of everything successful."

*By Jim Reger and David Irvine*

GUEST COMMENTATORS

Working with organizational cultures, the single most common request we get is how to build more trust and respect in the workplace. It is our experience that this is achieved through personal accountability—the ability to be counted on—which is the basis for personal integrity. Personal integrity leads to self-respect, respect for others who demonstrate integrity, and ultimately a respectful workplace. So in our view, personal integrity is the essence of building a successful culture of trust and respect.

As an engineer and inventor, Buckminster Fuller understood the importance of strength within a design. Engineers are accountable for designing structures capable of handling conditions up to a certain limit. In the engineering world, the margin between safety and disaster is known as "structural integrity." Similarly, our success as accountable people depends on our personal structural integrity. As the engineers of our own existence, our choices affect not only our own lives but also the lives of the people who rely on us.

By standing behind our promises and assuming a position of accountability, we begin to design a life of personal structural integrity. With this integrity as our foundation, our work and service in our families, organizations, and communities will be rock solid. However, just as you could never design and build a structure to handle any condition, personal structural integrity will always have its limits. Because integrity is not rigid, but instead strong and flexible and adaptable to life's changing circumstances, personal structural integrity can meet almost any test.

Integrity comes from the word "integer," which means wholeness, integration, and completeness. Being integrated is a necessary condition for self-respect, and self-respect is the basis for creating a respectful environment. Integrity means having clear, explicit principles and doing what you say you're going to do. It's about being honest with yourself and others. Integrity is deeply personal and, therefore, deeply applicable to all areas of life.

Integrity has everything to do with your success as a leader. Leadership—the capacity to elicit the commitment of others—is about presence, not position. Now more than ever, power, purpose, and privilege no longer reside at the top of an organization; they potentially live at every level. Great leadership cannot be reduced to techniques or tools or titles. While you may be promoted to being a boss, you don't get promoted to being a leader.

You have to earn the right to be called a leader. Great leadership comes from the identity and integrity of the leader. Authentic leadership is achieved through the power of presence, which comes from being an integrated human being, a person of integrity. Integrity is, indeed, the essence of everything successful.

DAVID IRVINE and JIM REGER co-founded the Newport Institute for Authentic Living, whose focus is on building authentic and accountable cultures of trust and respect that inspire and unleash greatness. They have co-authored two books on authentic leadership: *The Authentic Leader, It's About PRESENCE, Not Position*, and *Bridges of Trust, Making Accountability Authentic*.

DAVID IRVINE is one of Canada's most respected voices on leadership and organizational culture and his work has contributed to the building of accountable, vital, and engaged organizations across North America. David can be reached at: www .davidirvine.com.

JIM REGER's passion and commitment for facilitating powerful and effective change is evident in his work, which is focused on assisting entrepreneurial leaders in creating and building authentic lives and cultures. Jim can be reached at: www .regergroup.com.

△ THE LAST PUBLIC STATEMENT BUCKMINSTER FULLER MADE was *hold on to your personal integrity.* He made that declaration in response to a question asking what (after all these years of awards, famous inventions, books, accolades and other achievements) was the most important thing to him. His response followed his often-quoted remark, *only integrity is going to count.*

Bucky was so certain that integrity was the key to everything, that he named his beloved sailing schooner (the one seemingly extravagant material possession that he allowed himself) *Integrity.* He also made certain that his personal integrity was intact regardless of what other people thought or believed.

Bucky's definition of integrity is structural. Anything that has integrity holds its shape regardless of external circumstances. Success demands such a rigorous accountability. First, one must determine her or his "shape."

You need to know who you are, to have core values and to work at maintaining your internal and external integrity as much as possible. Then, like Bucky, you will have a life that you experience as genuinely successful.

You will also find yourself living your dream, making numerous contributions to others and feeling genuinely fulfilled. You will also find your "shape"/integrity challenged by external conditions and other people. Success will require that you do your best to maintain your shape just as Bucky and all the great teachers and leaders have done and are continuing to do.

And your success may not look like others would have it. You may be perfectly content living a simple, sustainable life rather than pursuing the "American Dream." You may appreciate walking and using public transportation rather than owning an automobile. And you may prefer spending your time with family and friends rather than chasing after the next big career promotion.

We all have to make such life decisions, but they are easier when you come from the context of maintaining your personal integrity—knowing that you are "holding your shape." Then, success is something that you choose rather than something that is imposed upon you by society, governments, parents, friends, or corporations.

**1.4** "Something hit me very hard once, thinking about what one little man could do. Think of the *Queen Mary*—the whole ship goes by and then comes the rudder. And there's a tiny thing at the edge of the rudder called a trim tab. It's a miniature rudder. Just moving the little trim tab builds a low pressure that pulls the rudder around. Takes almost no effort at all. So I said that the little individual can be a trim tab. Society thinks it's going right by you, that it's left you altogether. But if you're doing dynamic things mentally, the fact is that you can just put your foot out like that and the whole big ship of state is going to go. So I said, call me 'Trim Tab.'"

*By Werner Erhard*

GUEST COMMENTATOR

What Bucky says here stands out for me because Bucky's words speak powerfully to something basic in all of us—the desire to make a difference, to have our lives matter. Bucky refused to be limited by the conventional wisdom that there is nothing one little individual can do to make a big difference. This notion led to a resignation that became a frame for living for many people. Yet Bucky came to see that in fact each person is capable of making a profound difference in their own life and in the lives of others. He saw and spoke that the individual, any individual, has the power to take a stand, and live from the stand that who they are and what they do can make a difference, and that by doing so they become a trim tab literally capable of turning the ship of life.

Thanks to his grandson, Jaime Snyder (who had taken the est Training), I had the opportunity to meet Bucky. At the time, Jaime was 20, I was 40, and Bucky was 80—Bucky liked the symmetry. As Bucky and I got to know each other, I heard what could inspire people. We invited Bucky to meet with graduates of the est Training and their friends and families, and between 1976 and 1979 somewhere close to 100,000 people came to hear Bucky in seven cities across the United States. One thing I've learned from having the privilege of interacting intimately with tens of thousands of people is that the hunger to make a difference and contribute is fundamental for us human beings. People are willing to take great risks when presented with an opportunity to make a difference. In a moment of crisis, Bucky discovered that in himself, and he found a way to speak words that allowed other people to get in touch with that passion for themselves. I watched people when Bucky spoke, and people heard him and were deeply moved and inspired to action by his words.

In 1979 a group of UCLA scientists published a study of graduates of the est Training, "Separate Realities: A Comparative Study of Estians, Psychoanalysands, and the Untreated." The study concluded that est graduates had a high degree of concern for others (higher than the two comparative groups in the study). In the events Bucky did for est graduates he spoke to that higher degree of concern for others, profoundly validating each person who heard him. The analogy of the trim tab to shift the course of a giant ocean liner was met with enthusiasm. Participants adopted it and immediately began applying it for themselves, choosing the difference they could personally make for themselves and for their communities. We founded The Hunger Project (committed to the end of hunger) with Bucky's participation based on these same principles that the individual makes the difference.

Sam Daley-Harris and Jeff Bridges are well-known individuals who have demonstrated Bucky's principles. When Sam attended a Hunger Project event in 1977 he was a music teacher in Florida. Sam was deeply moved by the idea that "the little individual" could make a difference; he declared that he would create his "own form of participation in ending hunger," and in doing so created what has become his lifetime commitment to ending global hunger. Sam founded Results, which became the largest and most effective grass roots lobbying organization in the United States for the end of global hunger. In the late 1980s Sam met Muhammad Yunus, founder of the Grameen Bank in Bangladesh, and worked in partnership with him using the network he had established with The Hunger Project to get the Grameen Bank established throughout developing nations. In 2006 the Nobel Peace Prize was awarded to Muhammad Yunus and the Grameen Bank "for their efforts to create economic and social development from below." Sam continues his work as Director of the Microcredit Summit Campaign.

Jeff Bridges is best known worldwide as an actor. In the early 1980s, he brought together entertainment industry leaders and Hunger Project volunteers to found the End Hunger Network with a commitment to bringing Americans together to end hunger in the United States. And he didn't stop there. In 2010, Jeff became the national spokesperson for the No Kid Hungry Campaign, dedicated to eradicating childhood hunger in America by 2015.

These are only examples of individuals whose names you may recognize, but there are many thousands of individuals whose names you would likely not recognize who have embraced Bucky's words and live their lives as Trim Tabs.

Bucky made it clear and accessible that every human being can make a difference. I honor him for his humanity, for his friendship, and for the legacy of possibility he leaves with us. I honor Bucky for having created in himself the humanity that let him say *"Call me Trim Tab"* in a way that every individual can continue to take actions that make a difference, putting their own feet out to turn the great ship.

WERNER ERHARD is an original thinker whose ideas have transformed the effectiveness and quality of life for millions of people and thousands of organizations. While popularly known for the est Training and the Forum, his models have been the source of new perspectives by thinkers and practitioners in fields as diverse as philosophy, business, education, psychotherapy, third world development, medicine, conflict resolution, and community building. He lectures widely, and has served as consultant to various corporations, foundations, and governmental agencies. Erhard was acknowledged in *Forbes Magazine*'s 40th Anniversary issue as one of the major contributors to modern management thinking, and is a recipient of the Mahatma Gandhi Humanitarian Award. www .wernererhard.net.

By *Justine Willis Toms*

GUEST COMMENTATOR

In the 1970s when my husband Michael and I were in the early years of our broadcasting work, we had the enormous pleasure of spending many hours with R. Buckminster Fuller. Even though he was quite small in stature, his dynamic energy would fill a room. He would call everyone, men and women, *"darling,"* and insist they call him Bucky.

I was too shy to attend the first interview New Dimensions had with Bucky, so the first time I met him was at a reception in his honor, hosted by artist Ruth Asawa, in San Francisco. In those days, when entering new surroundings filled with strangers, it was my strategy to find a corner in which to hide.

I was in my corner when Michael and Bucky, who were talking together on the other side of the room, turned to face me. Then to my astonishment and panic, Bucky proceeded to navigate around tables, chairs, and couches, unmistakably heading in my direction. Arriving in front of me, he put out his hand and said, *"I'm Bucky Fuller."*

I remember thinking how he needed no introduction; after all, everyone in the room knew who he was. But I was soon to learn how unassuming and humble he was. It was obvious he was not puffed up by his fame.

Then he made an extraordinary statement. He said that sitting down with Michael for the radio interview was an evolutionary event. I was amazed by this acknowledgement. It was my first inkling of the potential of our work to make a contribution to others. Fuller was concerned about sustainability and human survival and was optimistic about humanity's future. Now, he was finding a new generation to join him in this most worthy endeavor. It was at that point he became a mentor for us and our work.

Since then I've spent many hours transcribing the numerous radio dialogues we've had with him. I was also privileged to participate in the weekend events called "Being with Bucky," cosponsored by his grandson, Jaime Snyder, and New Dimensions.

During those events it was sometimes difficult to follow his explanations. He talked at the clip of a racehorse going for the finish line. When he got into the complexity of the math that made up his *Synergetics,* I would feel lost. But that didn't matter because he was a transmitter. By virtue of his energy and enthusiasm, and his utter confidence in my ability to grasp what he was talking about, he could bypass my overwhelmed mind, and pour understanding and hope directly into my heart.

One concept of Bucky's that has remained a touchstone for me over the years is that we can all be "trimtabs,"—that is we can play a role in changing the course of things. A trimtab is a small device that is part of the rudder mechanism, which plays a crucial role in controlling the direction of a ship or an airplane. The metaphor was so important to him that, *"Call Me Trimtab,"* serves as the epitaph on his gravestone.

In an interview with Barry Farrell published in *Playboy* in February 1972, Bucky said:

*"Something hit me very hard once, thinking about what one little man could do. Think of the* Queen Mary; *the whole ship goes by and then comes the rudder. And there's a tiny thing at the edge of the rudder called a trimtab.*

*It's a miniature rudder. Just moving the little trimtab builds a low pressure that pulls the rudder around. Takes almost no effort at all. So I said that the little individual can be a trimtab. Society thinks it's going right by you, that it's left you altogether. But if you're doing dynamic things mentally, the fact is that you can just put your foot out like that, and the whole big ship of state is going to go."*

Farrell also quoted him as having said at the Buckminster Fuller Institute,

> "When I thought about steering the course of the 'Spaceship Earth' and all of humanity, I saw most people trying to turn the boat by pushing the bow around. I saw that by being all the way at the tail of the ship, by just kicking my foot to one side or the other, I could create the 'low pressure' which would turn the whole ship."

To understand this phenomenon, imagine a large ocean-going ship. To turn this enormous vessel in a new direction one must first adjust the trimtab, a miniscule rudder that runs the length of the larger rudder; once the trimtab is turned, the larger rudder follows. In fact, there are no mechanics yet devised that could turn the large rudder against the momentum of such a massive vessel without it breaking off. Only by first applying pressure to the trimtab will the larger rudder even begin to move, thereby changing the direction of the ship.

We might think of the planet Earth as a vessel moving through stormy seas. Each of us can be a trimtab by acting on our commitment to serve the greater good with our gifts, talents, experience, and confidence that we do make a difference.

Bucky was all about doing more with less. In the face of all that needs doing, I remind myself that I don't need to do everything; however, I do need to do something. So, I apply the trimtab factor to find what I can do that takes the least amount of effort, which will give me the most leverage. It is the way nature works: the way rivers flow; the way the wind blows. I don't have to feel overwhelmed by pushing against the momentum of my own life or even that of society. I need to find the trimtab that is mine, apply it and trust the difference it makes to the whole.

Whenever Bucky began a conversation with a new acquaintance, or even a group of people, he'd have to set the context from which all else would flow. This would include the entire history of civilization, and could take several hours or even several days. The last time we sat down with Bucky for our radio series, the format had been squeezed from four hours to one hour.

We were a bit worried how this conversation would go, knowing that Bucky would, no doubt, be setting the context. Bucky's book *Critical Path* was just off the press, and sure enough he launched into the history of civilization. After about forty minutes, he finished and there was a back and forth dialogue between him and Michael covering many other subjects. After the interview Bucky smiled, put a fatherly hand on Michael's shoulder, and said, "*Michael, I got it down to forty minutes just for you.*"

It was our deepest pleasure knowing and being with Bucky through those last years of his life. His ideas and hope for humanity and his confidence that each of us is a trimtab for good lives on.

JUSTINE WILLIS TOMS is co-founder, host, and managing producer of New Dimensions Radio/Media, which has been broadcasting continuously since 1973, producing and distributing life-affirming, socially significant, and spiritually relevant programming throughout the world. She is the author of *Small Pleasures: Finding Grace in a Chaotic World*. Learn more about her work at www.newdimensions.org.

⟁ "CALL ME TRIMTAB" IS THE INSCRIPTION BUCKMINSTER Fuller instructed be carved on the headstone of his grave, and these three words encapsulate how he was able to achieve so much in one short lifetime. The trimtab is a small rudder that changes the course of a ship with very little effort, and it served Bucky as a symbol of what "*the little man*" can accomplish.

Some cultures call this phenomenon leverage. The Buddhists refer to it as skillful means. It's the process of doing more with less that Bucky championed throughout his "56-Year Experiment" to determine and document what one individual could accomplish.

Regardless of the label, trimtab is how every person's action can make the most difference. It's also how "we the people" will redirect the path of human evolution from our current state in which we are dominated by war, competition, scarcity, and fear to our natural state of peace, cooperation, abundance, and love. And it's a tool that anyone can use to create the life they desire and *"a world that works for everyone."*

If you want to use the trimtab principle to create success with the least effort and resources, all you need to do is discover how to achieve more with less. And as you increase your "more with lessing," your project (and probably you) becomes more of a trimtab on many levels.

For example, I could (and often do) talk about Bucky with friends, family associates, and strangers, and I am more than willing to share what I understand with anyone who is interested and wants to learn. These one-on-one encounters are usually productive and enjoyable, but they're not very efficient. When I make similar presentations to groups of people, the event is more of a trimtab because I am achieving more with the same amount of effort. A group is also a trimtab for results because each person brings unique experience and insights to share. And the larger the group the greater the effect and the greater the trimtab factor.

Taking this analogy one step farther, by writing this book I again increase the effect as the book can reach many more people than I can in my personal engagements. I can continue that trimtab effect by recording my presentations and sharing those recordings on the Internet as well as in hard copy DVDs or mp3s. Further increasing the trimtab effect, these artifacts will continue to be of use to others long after I have died.

So, I say to you—"Call me Trimtab." And I challenge you to look for ways that you can be a trimtab for positive change in your life and on behalf of all sentient beings. That's the only way we humans are going to survive and thrive as a species and as individuals.

This is also true on a global scale. Most people believe that governments, corporations and other organizations can successfully steer the great ship that Buckminster Fuller named Spaceship Earth, but nothing could be farther from the truth. Our planet currently has hundreds of government, corporate, and religious leaders trying to do that, and they are failing because their actions are akin to people all trying to push a huge ocean liner from different places along its hull. Very few people understand that as Bucky explained, "*We need one captain, and that captain must have as much timely data as possible in order to make the most effective decisions regarding the course of our ship (global society).*" No single human or group of people can perform this duty because it is simply too enormous and no person can act from a completely unbiased perspective. No matter how noble our intentions, we all have prejudices.

Bucky often pointed out that only an impartial computer system can make the most effective, efficient decisions on behalf of all life. Once the computer makes these decisions, they can be implemented by a single captain or group of appointed officials who are chosen to do that job because they have excellent administrative skills in a particular area.

Such "leaders" are exceptional administrators who work in a position because of their skill and not as a result of the money they used to win an election. In other words, they are paid administrators not the elected politicians that got us into our current mess and who are in fact obsolete players working in an obsolete system called politics.

Bucky spent most of his adult life advocating that each of us follow what he had learned and become trimtabs for the causes and solutions that will create his vision of "*a world that works for everyone.*" It was his final message to us, and it remains carved in stone as a reminder to us all.

## 1.5 "We now have the resources, technology, and know-how to make of this world a 100% physical success."

*By Barbara Marx Hubbard*

Buckminster Fuller evolved my life in so many ways. The first was when I read his book *Utopia or Oblivion* in the 1960s. I was on a search for answers to two questions that arose in my mind after the U.S. dropped the atomic bombs on Japan. "What is the meaning of all our new powers that is good? What are positive images of the future equal to these powers?"

In my search I read through religion, philosophy, and technology. I found this quote from *Utopia or Oblivion*, and it changed my life and gave me the clue to the answer to my question. Bucky offered his Design Science Revolution as an answer. He had probed deeply into the way nature works. He said, *"Humanity is undergoing a viability test ..."* At that time he told us we had fifty years to demonstrate our viability ... or fail.

This is exactly what is happening. It's about fifty years since I read it. His insights galvanized me to try to understand how we could work with nature to transform the world.

I formed the Committee for the Future in Washington, and in 1971 we held the first conference called "Mankind in the Universe" at Southern Illinois University. Bucky was Scholar in Residence at that time. The conference was about our ability to develop a positive Earth/Space Human Development process.

I was speaking on the platform to discontented students dressed in costumes of protest. When I spoke of our potential to become a universal species, one of the students stood

VIEW: UNIVERSAL PERSPECTIVE     23

up and said: "Why would you want to send this disease—humanity—into the universe?"

I felt I had to defend humanity! An anger came over me. I said fiercely: "You, stand up now! How dare you criticize the whole human race that got you from caves to the moon? What are *you* doing that's good?"

He sat down!

The next day Tom Turner, Bucky's director of special projects came up to me and said that Bucky had an announcement to make. We were on the stage together. Everyone was attentive. Tom read from a note: "*Dr. Fuller recommends that Barbara Marx Hubbard run for president of the United States. It's about time that we have women carrying the positive options for the future into politics.*"

We were stunned. The audience stood up and cheered and signed a petition that I do this. Bucky felt that the idea of conscious evolution—our capacity to solve our problems and become a viable universal species—was right.

I didn't run then, but in 1983 he urged me again. I undertook an idea campaign to be selected as the vice presidential nominee by whomever would be nominated for president on the Democratic ticket. It was Walter Mondale. I was the other woman nominated along with Geraldine Ferraro.

I had had an appointment to see Bucky. I had been studying his *Critical Path*. I called him and said I wanted to know exactly what he felt were the most important issues for me to run on. "Were there any of his students I should talk with," I asked?

"*No, darling,*" he said, "*You must see me.*"

He died just before I could get to see him.

I did my best to put his ideas forward during the campaign. The key idea was that we need a new social function in the White House. At the time, I called it an Office for the Future and a "Peace Room" to be as sophisticated as the war room. Its purpose is to scan for, map, connect, and communicate what is working in the world to mobilize for coherent action. I felt that this would enhance the Design Science Revolution. It is about to be built now as Internet technology has advanced to the stage of being able to do just this.

The last thing I want to mention is the deepest. I had had a Christ experience in 1980. It had guided me to the New Testament. I was inspired by this Christ presence to write an Evolutionary Interpretation of the New Testament called the *Book of Cocreation*. It looked at the Gospels, Acts, Epistles, and Revelation with evolutionary eyes, realizing that our generation was gaining the power to do exactly what Jesus said we would.

He was a universal human—ourselves in the future. We are gaining powers we used to attribute to our gods. I could see that if we can evolve our own Christ consciousness (love) to guide our Christ-like capacities and powers brought by technology we can heal, we can produce in abundance, we can travel the Earth with the speed of light as holograms, we can do virgin births, etc. Then, we can actually fulfill Bucky's dream of "*a world that works for everyone.*"

Prior to his death, I sent Bucky a manuscript of the book I was writing. Later, I was scheduled to speak at the Annenberg Communication Center in Los Angeles with him, and he sent word down that he was "reading something" and that I was to speak to him alone. He wanted to see me afterward, in the Garden—alone.

He came down carrying my manuscript under his arms. We went into the Garden, and he closed the gate. He put his arms around me, touching his forehead to mine, saying, "*Darlin' there is only God, there is nothing but God.*"

He proceeded to tell me that one day he had been walking down a street in Chicago when he felt a light lift him off the pavement, and he heard the words, "*Bucky, you are to be a first mini-Christ on Earth. What you attest to is always true.*"

He told me that he had gone to the New Testament and had written a document almost identical to my *Book of Cocreation*. Then, he hid it. *"I could not use the words Christ or God,"* he told me, *"because I am an engineer."* He said he had hid the manuscript of his New Testament writings, but now he wanted to support my writings. When I asked for an endorsement he wrote:

> *"There is no doubt in my mind that Barbara Hubbard, who founded the Committee for the Future and helped introduce the concept of futurism to society, is the best informed human now alive regarding that movement and the foresights that it has produced."*

What he meant, I believe, is that both of us had a Christ experience that revealed to us the evolution of humanity from a self-centered, creature human to a whole-centered, spirit-centered universal human that is at one with the Processes of Creation and Nature.

When he put his forehead to mine I think he zapped me with the Design Science Revolution, as he did to so many others. He was and is a true light in our lives for the evolution of humanity.

BARBARA MARX HUBBARD is the founder of the Foundation for Conscious Evolution. You can learn more about her work at her website www.evolve.org.

THIS SIMPLE STATEMENT REPRESENTS THE ESSENCE OF Buckminster Fuller's most famous campaigns and is the hallmark of almost all of his work. He focused much of his life on educating people to this basic truth. And he was constantly crusading for the success of all people by reminding his audiences that such a feat was not just some pie-in-the-sky possibility but rather a necessity if humankind is to survive and thrive.

Bucky would often recount that comprehensive success is now the single option available for the survival of humankind, and that it had to be the success of everybody or nobody. This paradigm shift from "you *or* me" to "you *and* me" is essential to his message and the welfare of future generations.

Although he also worked in and talked about unseen metaphysical realms, Bucky is known for championing the fact that we now have the ability to support all people at a higher standard of living than anyone currently experiences. He was making this seemingly outlandish statement in the late 1970s, and it is as true today as it was then.

There is enough to go around, and we are fortunate that Bucky discovered that this unique paradigm shift would occur in his lifetime. During his 1930s stint at *Fortune Magazine*, he was able to access data on all global resources and to calculate 1976 as the year when technology would allow us to do so much more with fewer and fewer resources that there would be enough to support all life on Earth.

We now know that Bucky's prediction was correct for food, but it was also true for all resources, both physical and metaphysical. Still, fifty thousand people die of starvation every day of the year while food goes to waste in many regions of the world. This is not the "you *and* me" culture that Bucky said was necessary, and we need to make a drastic shift very quickly if we are to survive and thrive.

We have the *"resources, technology, and know-how,"* but we need to rapidly shift our use of them from weaponry to what Bucky labeled livingry. The distinction is quite simple. Weaponry-focused resources support death and destruction while livingry-centered resources support sustainable evolution and life. Livingry includes things like food, shelter, clothing, education, clean water, pure air, an infrastructure that supports the welfare of all sentient beings, and other life supporting activities.

This is the simple strategy that Bucky proposed again and again in his presentations and writing, and it can be applied to both global and individual activities. If there truly are enough resources to support everyone, each of us needs to begin acting appropriately. Greed and hoarding need to end immediately. War is obsolete as there is no need to fight over abundant resources. And people no longer need to work for a living or as Bucky used to say *"work to earn the right to live."*

We all have unique gifts that we can contribute to the whole of society and each other. In a world that is a *"100% physical success,"* we can each share our gifts freely with our neighbors be they across the street or on the other side of the planet.

# 1.6 "Love is metaphysical gravity."

*By Gary Zukav*

GUEST COMMENTATOR

Bucky said, *"Love is metaphysical gravity."* I agree. What else could it be? Without gravity you would float like an astronaut in a spacecraft. Up and down would mean nothing to you. Your slightest motion would send you tumbling head over feet or rolling uncontrollably. If you pushed hard against a wall, you would shoot backward fast until you hit another wall. If the lights in the spacecraft went out, you would have no way at all of orienting yourself.

Without love the same thing happens. Every experience of anger, jealousy, resentment, and fear sends you spinning out of control. You have no way of knowing up from down except what your anger shows you, and it always shows you that you are right and someone else is wrong, that you are a victim and someone else is a villain. The more you act in anger, jealousy, resentment, or fear, the more painful consequences you create. You careen helplessly, spinning, rolling, hitting walls you can't avoid and colliding with others.

Love grounds you. It orients you. Love brings your awareness to others and yourself. Love opens your mind and heart to others and yourself. Love settles you and gives you balance. When you choose to become sensitive and caring instead of frightened and selfish, your anger turns to appreciation, your jealousy to gratitude, and your resentment to caring. You cannot loose your orientation: When your deeds harm others, you are in fear, and when you create harmony, cooperation, sharing, and reverence for Life, you are in love. The ground beneath you is always solid.

That is why mystics say that only love is real. They also say that pursuing what is not real brings only pain. Now we are all becoming able to see for ourselves that we are parts of a larger fabric of Life and experience for ourselves what Bucky saw so clearly: Gravity calls you to the Earth. Love calls you to Life. And they always will.

GARY ZUKAV is author of *The Seat of the Soul* and *Spiritual Partnership: The Journey to Authentic Power*. His website is www.seatofthesoul.com.

BUCKMINSTER FULLER OFTEN SPOKE OF LOVE, BUT HIS definition was far from our romantic idea of the concept. For Bucky, love was all pervasive and inclusive of everyone and everything. Thus, the term *"metaphysical gravity."*

As far as we know, gravity is a universal cosmic principle. It exists everywhere and cannot be escaped while operating in physical reality, and in the realm of the metaphysical (things that cannot be experienced by our physical senses), the same is true of love. It pervades everything. It is the cosmic glue that binds us into a universal oneness that is now reemerging as more needed and relevant than ever on Spaceship Earth.

Another of Bucky's more poetic explanations of love is,

*"Love is omni-inclusive,*

*Progressively exquisite,*

*Understanding and tender*

*And compassionately attuned*

*To other than self."*

Rather than saying all encompassing, the poetic and scientific Bucky coined the term *"omni-inclusive"* to describe the limitlessness of love. This fits nicely with his definition of love as *"metaphysical gravity."* In both instances, Bucky reminds us that love is everywhere and infers that all we need to do is be present and feel to know true love in its purest form.

Bucky goes on to give voice to something beyond words when he says that love is *"progressively exquisite."* In other words, as time passes, love becomes increasingly delicate and beautiful. This insight could only come from a person who has experienced love over many years or remembered lifetimes, and Bucky wrote this poetic verse late in his life. From personal experience he knew that true love expands out over time like a spider's web.

He also recognized that true love is tender and founded on a compassion that can be experienced as a frequency attunement or focus on someone or something other than oneself. In other words, true love appears when a person's attention is on the welfare of others in an egoless state. Generally, we experience such a state for only for short periods of time even though we strive to maintain it on a continuous basis.

Bucky was able to sustain this unique state more than most because he followed a discipline of looking for what needed to be done that was not being attended to and doing that. Such an unselfish attitude is fundamental to giving and receiving the love we all seek. By giving his gifts freely, Bucky became a living demonstration of love in action, and because of his public openness millions of people have modeled their lives after his.

### 1.7 "Nature always knows what to do when it takes over after humans have signed off. Nature never vacillates in its instantaneous decisions."

*By Jack Elias*

I love this Bucky quote because it points out a simple and profound truth that always goes tragically unexamined—nature is instantaneous in its function. If we contemplated this reality and let a felt sense of this reality arise in us, we would quickly be free. I learned to focus on this 'nature of nature' from my Buddhist teachers, Shunryo Suzuki Roshi and Chogyam Trungpa Rinpoche. The teaching is, 'Let go and trust your Big Mind.' So, I think of Bucky as a natural Buddhist.

Big Mind is all-inclusive—appreciating Big Mind and contemplating Big Mind means to recognize that we obviously arise from Nature, live in Nature, and our substance is of Nature, known and unknown, seen and unseen. It follows that our function is identical to Nature's function—inseparable from it—and it makes sense to synchronize ourselves with our Nature Self.

This is where it gets interesting. We seem to be able to separate from that which cannot be separated from! How is this possible? It is possible only in dream, only in hallucination, only in trance.

I came upon hypnosis and neuro-linguistic programming (NLP) after studying and practicing Buddhism for fifteen years. I experienced what was valuable in hypnosis and NLP to be bits and pieces of Buddhist wisdom and technique. I

32     A  FULLER  VIEW       SIEDEN

developed a synthesis of hypnosis, NLP, and Buddhism I call Transpersonal Hypnosis and Hypnotherapy/NLP.

The view of this method is that we are in trance all the time—we think we live in the world but we live in our mind. Thinking this, hallucinating this, we overlook our actual activity and relationships. We can actually go in circles damaging ourselves and our environment and never recognize our mistakes and that we are creating own demise. We convincing (hypnotically) suggest to ourselves that it is someone else's fault. And then we create conflict, making things even worse.

When the pain or exhaustion is great enough, people may seek out help. They come for help believing change is going to be "hard," take a long time, and very likely that they are not up to it. I introduce what I consider to be a Bucky-Buddhist orientation to therapy. Change is effortless and instantaneous and we are full of all the robust resourcefulness we need— because our substance is Nature itself.

The only thing keeping us from experiencing effortless release from our problems is the hypnotic quality of language and of our thinking. It seems 'hard to change' when we think we have to battle our demons and our confusion. It is frustrating and fruitless because we are making mistaken effort in the wrong direction. Bucky also said,

> *"Everything in Universe is in motion, and everything in motion is always traveling in the direction of least resistance."*

We hypnotically ignore this quality of our own being to keep creating blockages, all the while telling ourselves we are trying to do better! Why do we suffer from these activities? Because the blockages we create in trying to make things better are blocking the spontaneous force of the universal energy moving through us in the direction of least resistance! The way to healing, clarity, and release from suffering is to stop fighting and grasping and to let go and let the flow of the universe wash us clean.

Buddha taught that the root of suffering is grasping based on the perception (hallucination) of self and other in threatening competition. Apparently he understood what Bucky understood—you can't fight Mother Nature without causing yourself suffering. She moves through us, as us, in the direction of least resistance (felt as joy and enthusiasm). The blocks we create are created out of our own energy—we block ourselves, our flow, by creating a counter flow, a grasping, and that creates pain and confusion.

Hypnotic suggestion creates confirming hypnotic hallucination. Because we hypnotically suggest to ourselves that we are doing the right thing, when we feel the pain and confusion, we see ourselves doing the right thing and we invest more energy in the blocking activity and so the vicious cycle begins.

We can turn this hypnotic grasping thinking around. We can end the false hallucination of warring self and other. As Bucky said,

> "Humanity is taking its final examination. We have come to an extraordinary moment when it doesn't have to be you or me. There is enough for all. We need not operate competitively any longer. If we succeed, it will be because of youth, truth and love."

JACK ELIAS (jack@FindingTrueMagic.com) is the author of *Finding True Magic: Transpersonal Hypnosis and Hypnotherapy/NLP*, and contributing author to *If I Were Your Daddy, This Is What You'd Learn*. He is the director of the Institute for Therapeutic Learning (http://FindingTrueMagic.com). Through on-site and distance learning programs, Jack teaches a unique synthesis of Eastern and Western perspectives on the nature of consciousness and communication to an international student body. In counseling and coaching (http://findingtruemagic.com/lucid-heart-therapy/), he presents simple yet powerful techniques for achieving one's highest personal and professional goals.

△ BUCKMINSTER FULLER NEVER CLAIMED THAT HE INVENTED anything new. Instead, he asserted that he was just observing Nature and following her lead. His famous geodesic dome is a perfect example of that strategy. It is an exact replica of the shell of viruses and other structures that occur naturally throughout our physical environment.

He also noted that Nature has and will continue to take over when humans have given up, been overwhelmed, or not known what to do. We only need look at the remnants of "lost" civilizations that have "disappeared" to see this principle in action.

In every part of the world we find the remains of cultures that have disappeared, and in every case Nature has taken over and returned the land to its natural state. Nature is relentless and persistent. She follows Winston Churchill's mandate to the British people during World War II when he chided them to "never, never, never, never give up."

Nature, however, does not need a reminder to never give up. She simply continues doing what needs to be done to maintain a sustainable balance and to support all life on Earth.

Unlike humans, she does not stop to hold a committee meeting when a crisis arises. She does not seek the expertise of anything or anyone other than what Bucky labeled "the divine mind always available in each individual." Nature takes immediate action in support of balance and sustainability, and her actions are always exactly what is needed—never using too many or too few resources.

Nature is the perfect administrator of all activities, and we humans would be wise to follow Bucky Fuller's strategy of continually mirroring Nature in all our activities and creations. That was how he came up with Synergetic Math. That system is based on the integrity of Nature's construction and triangulation rather than the manmade concept of math and construction based on unstable right angles. Nature builds with triangulated structures because they balance tension and compression internally.

Bucky recognized that we humans, like everything on Earth and throughout the cosmos, are part of Nature. He did his best to learn the laws Nature imposes on all things so that he could work within those laws rather than the laws made by men. He also respected the enormous power of Nature, and her innate desire to have everything (including humans) succeed. This is evident in his statement,

*"Nature is trying very hard to make us succeed, but Nature does not depend on us. We are not the only experiment."*

He understood that a perfectly balanced, sustainable system such as Nature would never depend on a single entity or site to solve a critical problem or fulfill a critical need. This is evident when we consider something as obvious as the human body. Nature designed all our important systems with a back up. We are designed with two eyes, two ears, and even two kidneys.

When Nature sets out to solve a problem or get a job done, she always has at least one back up in her design. She also moves quickly as is evident when grass and flowers quickly start springing up between the cracks of manmade pavement. And she never gives up in supporting the sustainable success of all creation. Accordingly, we would be wise to follow Bucky's lead and mirror Nature in all our activities.

# Universal Perspective of Gratitude and Equanimity

## Chapter Conclusion

**1.8** **"I don't have any favorite places or people. I love the whole show. A large number of beautiful people have taught me a great deal, and I am deeply indebted to them for their support."**

Daily Life on Earth is a "team sport," and it works best if we are on a "team" of people we love. After decades of looking to see what worked and what didn't, Buckminster Fuller realized that he (like all of us) was on a global team—the crew of the planet he named Spaceship Earth. He also came to understand that the *"whole show"* is an amazing series of relationship experiences, and that everybody living on Earth is, in fact, a player in the team sport "whole show."

After he attained this perspective, Bucky felt and often expressed great gratitude toward everybody and everything. Even in the last years of his eighty-eight year life when he was still busy writing, speaking and inventing, many people would be surprised to receive a personal response from him to a letter or phone call. He truly loved all people and was grateful for what they had taught him and contributed to his life, and he realized that he had a responsibility to do the same for others. This was particularly true of children who regarded him as the wise old grandfather and for whom he would do his best to make time.

Not one to ignore the "little individual," Bucky was supportive of many people and initiatives, and he did his best to love everyone equally. In the later years of his life, he said that he had no favorites because he had learned that we are all One, and his daily life reflected that creed.

In almost all his thoughts and actions Bucky was keenly aware and appreciative of the beauty and significance of every person, place, and thing. So, he really did not have favorites, and he treated all people and sentient beings with equal respect and admiration.

Eventually he began to inspire his fellow crewmembers on board Spaceship Earth by continually reminding us that he was simply an average individual doing what he saw needed to be done that was not being attended to. He would follow that statement with his conviction that those listening to him were capable of much more in their lives than he had ever achieved. By telling people that "learning experiences" (a.k.a. mistakes) are necessary to our evolution and growth, he would remind his audiences that the reason he was standing in front of the room speaking was that he had made more mistakes than anyone in attendance.

Life was an ongoing learning experience for Bucky, who was considered a failure for much of his life. Still, he was continually grateful for every experience that came his way and every person who helped him on his path even if that "help" appeared as a challenge.

His ongoing equanimity and eternal gratitude are two of the practices that led to his success as well as to the love and respect that was showered upon him during the last decades of his life. He accepted and was appreciative of everyone he encountered, and that view led to his being fondly called "Grandfather of the Universe" by thousands of his fellow Spaceship Earth crewmembers.

CHAPTER 2

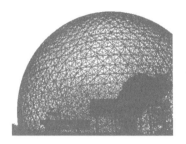

# PURPOSE

## HUMANS ON EARTH

## 2.1 "The most important thing about Spaceship Earth—an instruction book didn't come with it."

⚠ OUR HOME PLANET DID NOT COME WITH AN INSTRUCTION book, and Buckminster Fuller was one of the first people to try answering the question of how we can most effectively and efficiently function on Earth. In 1968 he wrote the classic book *Operating Manual for Spaceship Earth*.

Bucky realized that the Earth and all its inhabitants are constantly changing, and we must continually revise the way we function on our Spaceship in order to survive and thrive. Currently, we're at a critical moment that Bucky defined as *"humanity's final examination,"* and we still don't seem to have realized that Nature has provided us with a perfect operating system. It's all there and available to us, if only we would follow the clear path that Nature / Universe / God / Great Spirit / Higher Power / Greater Intelligence uses to create and manage everything.

Most humans believe that we know better. We think that we can impose our will upon Nature. We believe that it's a question of us versus them, man versus Nature or survival of the fittest. Nothing is farther from the truth that Bucky and other wise people have discovered.

If we want to survive and succeed as a species, we can no longer continue functioning as our ancestors did. We know that their unsustainable ways are not applicable in a highly industrialized and more technically savvy culture. We have the knowledge, and we need to put it to use on behalf of all sentient beings.

Today's *Operating Manual for Spaceship Earth* would tell us that we must include all people and living beings in the formula for success. It would again emphasize that in 1976 we entered a period of abundant resources in which there is enough to support everyone. Within that context, war, poverty, starvation, and politics are now obsolete.

The primary means for manifesting this already present paradigm and saving the human (and many other) species from extinction is a shift from weaponry to livingry. This transformation is essential for the successful operation of Spaceship Earth today.

It becomes an increasingly viable shift as more Spaceship Earth crewmembers become aware that humans have created this situation and as more of us take action to solve our critical problems. Those same first responder / early adopter crewmembers are also the people who will remain calm and focused on our collective well-being as our environment becomes increasingly chaotic and fear-based.

Bucky was one such early adopter, and he created a template for effective, efficient human action that is more vital today than it was when he was alive. He determined and documented what one individual can accomplish on behalf of all humankind that cannot be achieved by any government, corporation, or other institution—no matter how large or powerful.

We can all learn from his example of what the *"little individual / average healthy man"* can achieve and go well beyond what he was able to accomplish because of the "Operating Manual" that he left us. That Manual was more than the thin book he published in 1968. It is, in fact, his well-documented life, which is a model for living as a conscious human and employing technology both physical and metaphysical to create a sustainable environment that supports all life.

The fact that today's ever expanding technology allows one individual to make a greater impact than the largest, most powerful organization or institution is a critical element of a modern *Operating Manual for Spaceship Earth*. We can now each follow Nature's basic instructions in creating a sustainable, abundant planet for all sentient beings. We can write our own Operating Manual based on love, cooperation, peace, joy, union, and the success of each and every person.

This modern approach to life that melds technology and compassion is critical. It would have probably been in the first chapter of a written instruction book for our planet—if we had been provided with one.

## 2.2 "The most special thing about me is that I am an average man."

*By Robert White*

GUEST COMMENTATOR

Clearly these are not Bucky's most intellectually challenging words. Yet their meaning for me was profound and useful in my life journey. They could play the same role in yours.

1980 was my seventh year living and working in Japan when the Tokyo American Club hosted Buckminster Fuller's speech. I had read the books, marveled at the dome, and been fascinated by the Dymaxion car at the Chicago Museum of Science and Industry. We used John Denver's birthday gift song for Bucky, *What One Man Can Do*, in our seminars, and I'd often thought about how wonderful it was to have a song written for you. John did write a song for me after his experience of our seminar eight years later ... but that's a different story.

Bucky was one of my heroes, and I might have been the first person to sign up for the event. I arrived early, notebook in hand. I strategically selected my seat. For me, he was a rock star and I was experiencing a kind of giddy excitement while attempting to remain a cool Tokyo businessman.

The introduction was what you would expect. Lengthy and laudatory—a fitting tribute to a man who had lived his life in service to humanity. Then this tiny, frail giant slowly walked to the stage while we gave him a standing ovation. His first words after all that praise were:

*"The most special thing about me is that I am an average man. I say that as a challenge to any limitation you may have accepted for your life."*

I was shocked and barely remember the balance of a one-hour speech. My business success and lots of therapy had not been adequate to erase a history of feeling "less than." I felt I needed to stretch in order to get up to average and here was one of my heroes saying firmly and sincerely that he was just that: average. It was one of those moments where it's possible to have one's worldview altered and it was happening to me. Something shifted within me and it has made all the difference in my life.

Bucky stole my best excuse for being and doing less than my potential. While I was comfortable knowing that I would not approach his contribution in the world, I could no longer use the "I'm not enough" excuse for doing what was possible for me in my family, my company, and my world. After all, Bucky was also average.

When I later grappled with general systems theory I stepped up to the intellectual struggle—especially when we were taught about the power of leverage points. Of course that was just a different version of Bucky's trimtab teachings. When John Denver and I traveled together to deliver the Windstar Foundation's Higher Ground event—where we played a modified version of the World Game—I was able to honor and celebrate Bucky's life—and mine.

Recently, someone asked me about those serving in the American military. "Where do we find these people?" I'd ask the same about Buckminster Fuller. I will be forever grateful.

ROBERT WHITE founded human potential training companies Lifespring, ARC International, and Extraordinary People. He's an executive coach and author of *Living an Extraordinary Life*. You may reach him and register for his free weekly e-zine "An Extraordinary Minute" at www.ExtraordinaryPeople.com.

⚠ THIS SIMPLE STATEMENT REPRESENTS THE ESSENCE OF Buckminster Fuller, and he shared it often in both writing and speaking because he wanted people to realize that he was not special. In reality, he was unique not because of his intellect or amazing metaphysical capabilities but because of his willingness to take action in support of what he felt was most important, and that was usually the welfare of all life on Spaceship Earth.

We are all "average." We're simply humans operating out of physical bodies on Earth. We're also unique and capable of accomplishing much more than we believe ourselves to be. We can accomplish anything we set our minds to, and Bucky proved that throughout his life and especially during this "56 Year Experiment" to determine and document what the *little individual, average man*" could accomplish.

Bucky's accomplishments are well beyond what most people achieve, but by reminding us again and again that he was an *"average man,"* he reiterates the fact that anyone can accomplish what he did and more. This is particularly true today when our ever-expanding technology allows each of us to do so much more with fewer resources.

When the *"average man"* Bucky Fuller was conducting his experiment, a phone was a static device that connected people via wires and a computer was an expensive tool that was available only to a few. Today, most average people have access to these and other tools that allow us to be much more effective and efficient in finding and sharing our gifts.

Bucky realized that every person is a gift and has skills that she can share with others. That is the "average" that he recognized and championed throughout his life, and in an abundant society we no longer have to put a price on that service. All average people can now give of themselves freely knowing that if what they share is genuinely from the heart with no strings attached, Nature / Universe / Source / Greater Intelligence will support them in all aspects of their lives.

This is the challenge that "we the people / average women and men" face today. We must confront our fears and a society dominated by a few individuals who want us to believe that there is not enough to support all life and that we must fight for our "share." Nothing could be further from the truth, and Bucky was one of the first people to recognize this paradigm shift and campaign for all of us to wake up to our natural birthright.

We "average" people are the ones who will humbly create Bucky's vision of *"a world that works for everyone."* And we'll do it with great compassion, joy, love, grace, and integrity. That's how we must behave if we are to pass the period Bucky labeled our *"final examination."*

Average people doing what needs to be done with great integrity and imagination. That's our formula for success during this unique period of evolution.

## 2.3 "It is not for me to change you. The question is, how can I be of service to you without diminishing your degrees of freedom?"

*By Greg Voisen*

Bucky Fuller's quote has significant meaning for me, especially because of my chosen vocation and calling. I am a transformational leader in the personal growth, spirituality, and wellness space. I have interviewed hundreds of personal growth / mastery authors, and the above quote resonates with my lifelong personal passion for learning and for teaching others.

Our role as teachers, learners, and mentors is not to change others but to nurture the inherent wisdom that resides within and to allow this intelligence to illuminate from our souls. In being of service our opportunity is to help others find the magic elixir, the flame of passion that resides within, and to encourage the release and expression of this intelligence. The learners transform emotionally, spiritually, intellectually, and psychologically, and they are immediately released to be free from any bonds they may have placed on themselves inhibiting their unique expression and an opportunity for personal transformation and growth.

Awareness, authenticity, and the genuine expression as an alive and free soul is the most beautiful aspect of all human life. No barriers, no bars or obstacles can prohibit one from his or her unique expression. This newfound freedom becomes a critical link in the chain of all humanity, especially as it relates to being courageous, compassionate, caring, and kind. Manifesting these qualities are what heals our souls and the collective consciousness of the fellow travelers on this planet.

The philosophies and ideas of Bucky Fuller were first unveiled to me by author and speaker Mark Victor Hansen. I remember attending a meeting in the early '80s, listening to Mark speak so enthusiastically about how Bucky transformed his life through his ideas and teachings.

Mark had the privilege of assisting Bucky Fuller at the peak of his career. He dedicated a significant segment of his speech to discussing the inventions of Bucky Fuller, such as the Geodesic Dome and the Dymaxion automobile. I was personally transformed and influenced by Mark's amazing stories and today, some thirty years later, I frequently speak about the history, lessons, and ideas that Bucky gifted to humanity.

As Bucky stated, we are here not to diminish the freedom of another nor are we here to change people. Our role is to be the guiding light illuminating the way for others to find authentic expression, and in so doing make a contribution to the greater good for all humanity.

GREG VOISEN is an author, speaker, blogger, and the founder of Inside Personal Growth. Inside Personal Growth is dedicated to providing listeners with engaging, dynamic, and transformational interviews with authors in the personal growth, mastery, spirituality, and wellness genres. To learn more about Greg Voisen and his dedication to spreading the message of human potential, please visit his website at www.InsidePersonalGrowth .com.

△ SERVICE TO OTHERS WAS A CRITICAL ASPECT OF BUCKMINSTER Fuller's life. He was born into a family of ministers and lawyers at a time when ministers and lawyers were respected, and his father was a global importer at a time when few people traveled farther from home than they could ride on a horse in one day. Thus, Bucky grew up within an environment where contributing to others and a global perspective were important even though few of his peers or elders appreciated these concepts.

Bucky eventually melded these two ideas to create his vision of all resources being used to support all humankind.

That vision morphed into the idea of *"a world that works for everyone."* In this abundant, sustainable world, we each need to be of service to others without attempting to change them.

Within this context, it is up to each of us to understand that nothing we do or say will change another person. They will do what they will do regardless of our efforts. Trying to change another person usually escalates the problem for everyone involved.

Bucky understood that the only way to create change in another is to transform their environment. One of the easiest ways to change another person's behavior is to change your own. As Gandhi said, "Be the change."

Another way to change another person's environment is through artifacts. You can either create new environment-changing artifacts such as a geodesic dome or this book, or you can share existing artifacts such as this book or other websites with another person. In either case, the person's environment will have changed because we can't learn less. In other words, when we are exposed to something, we learn from it, and we can't unlearn that information. Thus, a person's environment (internal and external) is changed with the inclusion of new "information."

They remain as free as they were prior to learning this new thing, but that freedom has altered. They now possess more refined yet expanded views, which usually prevents them from taking actions that will be harmful to themselves or other living beings. And, thus, the person who introduced the change to their environment has been of great service to them and increased their freedom.

**2.4** "Humanity is taking its final examination. We have come to an extraordinary moment when it doesn't have to be you or me. There is enough for all. We need not operate competitively any longer. If we succeed, it will be because of youth, truth, and love."

*By Hazel Henderson*

GUEST COMMENTATOR

I remember being with Bucky in Philadelphia for a World Game session, organized by Medard Gabel, who is still furthering Bucky's vision with his Big Picture presentations at the United Nations and worldwide. Bucky and I both left to catch the same USAir flight up to Boston for another group of his followers led by John Todd, who with Nancy and Jack Todd created the New Alchemy Institute. Among their famous alumni are Gary Hirshberg, founder of Stonyfield Farms and their yogurt products, and Sim Van der Ryn, pioneer green architect.

Bucky and I were lucky enough to find seats together, and I was awed with the privilege of having one hour in his august company. Intuitively attuned to Bucky's daring naïveté and technical brilliance, I was imbibing his expansive consciousness almost through the pores of my skin.

Looking closely at Bucky, I realized he was tired. So I took his outstretched hand, and we both agreed to take a nap. We awoke as our plane landed in Boston. Bucky, fresh as a daisy, and I, still bemused at my good fortune, arrived at the next conference. I sat in the front row and drank in another virtuoso Bucky presentation.

Since then, I often use Bucky's deepest statement on education for our time of whole-system transition on planet Earth:

"*Humanity is taking its final examination. We have come to an extraordinary moment when it doesn't have to be you or me. There is enough for all. We need not operate competitively any longer. If we succeed, it will be because of youth, truth and love.*"

and its corollaries:

"*It is now highly feasible to take care of everybody on Earth at a higher standard of living than any have ever known. It no longer has to be you or me. Selfishness is unnecessary. War is obsolete. It is a matter of converting the high technology from weaponry to livingry.*"

"*We are called to be architects of the future, not its victims. The challenge is to make the world work for 100% of humanity in the shortest possible time, with spontaneous cooperation and without ecological damage or disadvantage of anyone. How can we make the world work for 100% of humanity in the shortest possible time through spontaneous cooperation without ecological damage or disadvantage to anyone?*"

For me, these truths led to my lifelong interest in technology and our levels of competence in understanding the planet's geology and biological mantle of life forms, including humans, and how we all share the same atmosphere, water, and climate.

I saw the great transition still underway from fossil-fueled industrialism to the wiser, cleaner, greener societies in my *Politics of the Solar Age* (1981). Bucky identified all these challenges of this transition and our new abilities to provide for 100% of the human family—by redesigning our cities, societies, cultures, and beliefs.

Most crucially, Bucky knew we would have to design our systems of money-creation and credit-allocation beyond their basis in scarcity, fear, and competition. These tasks are now at the top of the human agenda and even army generals are seeing war as futile and military approaches as less effective than diplomacy.

Recent financial crises have demonstrated the corruption and collapse of our global money circuits. The future lies not in cries to return to the gold standard but to acknowledge real forms of wealth beyond money: healthy, wise humans interacting with the planet's ecosystems, using the free daily photon shower from our sun as effectively as plants do in photosynthesis.

So, for me, Bucky Fuller lives on in my "Love Economy" research and in our Ethical Markets Green Transition Scoreboard®, our Transforming Finance initiative, the Calvert-Henderson Quality of Life Indicators, and our Beyond GDP work and surveys with Globescan.

Yes, Bucky. *"Love is that metaphysical gravity"* that keeps us all one human family. And I continue to *"Dare to be naïve."*

HAZEL HENDERSON, D.Sc.Hon., FRSA, is President and Founder of Ethical Markets Media (USA and Brazil). She is a futurist, evolutionary economist, and author of the award-winning *Ethical Markets: Growing the Green Economy* and many other books. Her editorials are syndicated by InterPress Service, and her articles appear in journals worldwide. She leads the Transforming Finance initiative, created the Green Transition Scoreboard®, tracking global private sector investment in green tech, and developed with Calvert Group the widely used alternative to GNP, the Calvert-Henderson Quality of Life Indicators. In 2010 she was honored as one of the "Top 100 Thought Leaders in Trustworthy Business Behavior 2010" by Trust Across America.

*By T. J. Mackey*

GUEST COMMENTATOR

Buckminster Fuller was a teacher and now I am a teacher. I work in a system of education that uses competition as its vehicle. If your only goal is to find out who is the best, then this system seems to make sense. But if you value something else then it would make sense to call this system into question.

Many feel that our education system is broken, or at least pretty far off track. Yet we often hear a cry to increase accountability. The way we do that, we are told, is more testing, hence more competition. Creating competition between students, between teachers, between schools, between districts, and even between states.

What if we stepped back for a minute and asked a different question? What if our schools were based on cooperation rather than competition? What if students had to work together to answer complex problems and not race to win a standardized exam?

The geese fly faster and farther working together to form their "V." The fish find safety working together as a school. Evolution welcomes efficiency.

Buckminster proved over and over again that the Earth can and will provide. Man must cooperate in order to share in it, but we must also provide for the needs of others. Hungry students don't learn well. Students who are not safe or are alone can't focus. Students consumed with chasing material gains or accolades don't aspire to new challenges. By meeting the needs of those around us we all move forward together. Schools can and should be teaching that, not competing for funding.

In my classrooms, we use inquiry. We listen for the questions, both teacher and student. We seek depth rather than the quick and easy information on the surface. Students figure out how to work together to reach their goals, not ones handed down from above. Classrooms can allow for creativity and tangential thinking if they are not slaves to deadlines. "Deadline" seems appropriate, because it often leads to a dead line of thinking when students are marched on to the next topic. Students can learn to listen to each other; they can learn to work to support each other too.

T. J. MACKEY is Middle School Head at Seattle Country Day School.

A WELCOME TO OUR FINAL EXAMINATION. YOU HAVE ALL THE resources to succeed, and so does everyone else taking this exam. There will be no grades. The time allotted for this test is as long as you need, or more correctly as long as you and all your fellow travelers on Spaceship Earth survive.

This is a "pass/fail" test, and everyone must pass if humanity is to continue on Earth. This exam is a "team sport," and every living person is on your team. Please pick up your pencil and begin. The clock is ticking!

Buckminster Fuller often pointed out the seemingly simple answer to this test when he explained that we must shift our resource usage from weaponry to livingry. In other words, we live in abundant Universe on an abundant planet, but we need to stop trying to kill each other and start feeding, housing, educating, and clothing people.

This is the mandate of *"youth, truth, and love"* that Bucky so adamantly championed. The "adults" of the world have made a mess of our tiny, fragile Spaceship Earth, using lies, half-truths, and fear to dominate and control. Some have made a small change in humanity's quest to destroy the planet and all life forms, but we need the majority of "we the people" to stand up for truth and love.

As Bucky found, the people who most care about this overriding issue are the "youth," no matter what their physical age. Although most are physically young in years and somewhat idealistic, Bucky also seemed like a youth for most of his life, and the younger generations flocked to his marathon "thinking out loud" lectures.

They saw the truth and passion in his words, and they respected the fact that he had "done his homework" and found that the human experiment could succeed. They bought into his vision of *"a world that works for everyone,"* and they were among the first who would stand with Bucky and *"dare to be naïve."*

Now is the moment for all of us to declare our youthfulness, for us to stand together as one people living on one planet that has the ability to support us all. Now is the time, and we are the people. We're all in this together, so we might as well get with the program and follow Nature's guidance. It's a universal issue that each of us gets to explore during our individual and cultural "final exam." As Bucky so often reminded us, *"The cosmic question has been asked. Are humans a worthwhile to Universe experiment?"*

Bucky said "yes," and I agree. Still, we have to work together to pass this test. The clock is ticking, so let's get started.

## 2.5 "All children are born geniuses. 999 out of every 1,000 are swiftly and inadvertently degeniused by adults."

*By Anna Beshlian*

GUEST COMMENTATOR

Recently, my Language Arts teacher read Antoine de Saint-Exupéry's *The Little Prince* to us, and I am struck by how closely related Bucky's quote is to the beginning scenes of the story. The narrator, as a young child, draws the silhouette of a boa constrictor eating an elephant and shows it to his parents. With just a glance and no hesitation, they identify it as a hat, and bluntly state that they see no resemblance whatsoever to a boa constrictor digesting an elephant.

The narrator's discouragement grows and eventually develops into self-doubt that pervades his mind and obliterates his dreams of being an artist. He instead becomes a pilot, flies across the Sahara desert, and crashes. His dreams of being an artist are rekindled when he meets the Little Prince in the middle of the desert; he understands the narrator's drawing of the boa constrictor eating an elephant. The narrator eventually travels with the Little Prince to different planets, each habitant being a different character symbolizing faults in adults' behavior and decisions.

If asked to comment on his quote, I imagine that Bucky would basically say—in his creatively cryptic way, of course—that older generations must let new generations grow up without any degrading aspects of life that "degenius(es)" them. I wonder what Bucky would have to say if he saw our modern-day society. Our society of adults, similar to the narrator's parents, is unknowingly responsible for gradually depleting the sense of genius innovation that the younger generation is born with.

Individuals able to gracefully transcend rules in our society are rare. As children mature, they are constantly bombarded with examples of proper behavior, limitations, and rules (first introduced by their very own parents) about how their genius innovation should work.

My own experience of being "degeniused" was prominent in my mid-childhood, when I attended my neighborhood public elementary school. While I received a good, solid education, it wasn't the inquiry-based type that my brain desired. Rather than explaining how or why things worked, facts were fed to me in textbooks, out of which I would frequently read, memorize the material, and spit back answers.

Questions that would jump-start my thinking process were rarely asked. I also highly disliked my teacher's responses when I would ask her a question that I considered vital to my understanding of that topic.

If her answer was no, I would then proceed to ask why she said no. Her second answer would be something along the lines of "that's just the way it is" or "because whoever wrote the story wanted it to be that way."

Of course, perhaps "that's just the way it is" is a legitimate answer, because maybe there is no explanation for how something is. And maybe whoever wrote the story indeed "wanted it to be that way" for a mysterious reason unknown to the general public. Still, their behavior astounded and angered me, and I was left alone in the dark corner of my doubtful mind, unable to see her reasoning.

How do we solve degeniusing? Perhaps we can't, as most of us prefer to be in control of our lives, and to harmonize with those surrounding us. Maybe children need to be "degeniused by adults," in order to function well in our society and be successful.

Or maybe introducing a little creativity into our lives never hurt anyone, and we need to learn to accept ourselves for who we are. Let's not change anyone's personal identity, nor let anyone change ours.

ANNA BESHLIAN is in the eighth grade at Seattle Country Day School.

*By Stephen Garrett*

GUEST COMMENTATOR

I have always admired Buckminster Fuller's genius, his creative force and productivity. I chose this quotation because by being himself he demonstrated that there is a way to maintain the genius throughout a lifetime.

I resonate with this quotation, as I have long "known" that we are all magnificent potential waiting to get out—all that has happened is life! We are often reminded about all our apparent deficiencies, all that we are in negative ways. We are seldom reminded of our natural genius or greatness. Bucky is right—we get degeniused!

A study done several years ago with two-year-old children revealed the following very shocking information. Researchers spent twenty-four hours with the average two-year-old and monitored two things:

1) Negative comments the child received, and
2) Positive affirmations about the child.

Here are the results! Negatives in twenty-four hours: 432. Positives in the same period: thirty-two.

That is a 13:1 ratio of negatives over positives! Any wonder self-esteem is our nation's number one problem in my estimation. So get this—a child sixteen years of age has heard approximately two-and-a-half-million times how bad, wrong, not good enough negatives about themselves and only 186,880 positives!

This is exactly what Bucky is speaking about. The genius gets drowned in a sea of negatives; lost in the illusion of these negatives, is it any wonder most of us are not living a great life to our full and magnificent potential. Is it any wonder that we are all seemingly stuck in the matrix?

What I have discovered over the years of my own personal growth is that these negatives can be transformed back into our natural genius with simple and profound techniques that require only a willing client and a deeply loving therapist. With some therapeutic skill and technique and a lot of "heart-art," the uncovering of the dormant genius can be simply and gracefully accomplished. The negative illusions that were so unceremoniously implanted in us can be removed, freeing the real one we are to live a life of freedom, passion, and love.

Once the genius of each of us has emerged and we are once again able to live as that genius there is no limit to what we can create on the planet. This is just the point Bucky is making in the quotation—*"Everyone is born a genius, but the process of living de-geniuses them,"* and more, the point of his life.

Gandhi once said "Be the change." Perhaps we can combine Bucky and Gandhi and use this quotation as a mantra—"Be the genius you were born as." If we can get the job of reclaiming the genius in each of us done there is nothing we cannot accomplish.

So I would recommend that we all take on the fun and playfulness of reclaiming our own natural genius. There are some way-showers out there that are the real thing and will gladly help others remember who they are at their core, at their source and essence. Find them, be with them as long as you need, and then set out on your own to pay the genius of you forward. It's very much spiritual anarchy—come play. It is so much fun!

STEPHEN GARRETT is an author, trainer, coach, and guide. In 1988, he woke up from a deep sleep called the North American Dream and has since been devoting his time, dynamic energy, and life to helping people wake up to their own personal power and essence and learn to live their lives based on their heart's wisdom. He has written two books, *Men Read This—A Spiritual Guide for the Regular Guy* and *Monks Without A Church—Life Beyond Religion*. He owns and operates Just Alive Consulting with his wife Sonora, and they offer retreats and yearlong coaching programs designed to support others in reclaiming their own genius and then living from it. You can find him at www .stephengarrettnow.com

△ BUCKMINSTER FULLER DEVOTED A GREAT DEAL OF HIS young life resisting being "degeniused," and in his early adulthood he "re-geniused" himself. He was fortunate to have been raised in a family that valued education, even though they tended toward traditional New England values that dominate the way people lived and thought.

Because he had essentially been an undiagnosed blind child for the first five years of his life, he was not subject to many of the usual factors that tend to "degenius" most of us. It was only after getting glasses that he could see clearly and was expected to follow the mandates of the adults who wanted all children to be seen and not heard and to follow a strict path toward becoming "good citizens."

That path also tends to eliminate thinking for oneself, and Bucky continually rebelled at that notion. Without the benefit of sight during the first five years of his life, he had developed extremely efficient use of his other senses, and they connected him more deeply than most to Nature. He would attempt to taste, touch, smell and hear everything in the semi-rural environment of his home and the extremely rustic environment of Bear Island.

Yet, despite that advantage, he still fell into the trap of having to "become responsible" and follow the dictates of other people. That led him to try attending Harvard where he was expelled twice for not following the rules and to take several jobs that did not suit his freethinking nor his desire to experiment and contribute to the world.

In all those seeming failures, Bucky learned much about life and his role in the larger scheme of things. He also began to uncover the broad perspective that would eventually result in his vision of a viable "world that works for everyone." Still, when he began to share that vision, he was seen as a crackpot, pie-in-the-sky dreamer rather than a wise prophet of the future.

In the later years of his life, when he was finally recognized as a genius with an important message, the primary audience were young people who, because of their age and the era in which they were raised, were more open minded than previous generations. They recognized a kindred spirit and wise elder in Bucky, and they intently followed his advice and his re-geniused life path.

Still, Bucky's awakening was the result of many agonizing "learning experiences" (mistakes) including the painful death of his four-year-old daughter Alexandra in his arms. Eventually, he realized that society was and still is attempting to turn children into "productive citizens," which essentially means training them to be cogs in an industrial complex that focuses on increased consumption, profits, greed, and war.

Bucky realized that learning to think for himself and trust his personal experience over the words of others was far more important than earning a living or, as he often called the process, "earning the right to live." He recognized that our society is educating natural curiosity, comprehensive nature, cooperation, and talents out of our children in order to make them robotic, non-thinking, expendable pawns in the fear-based production machine we call developed modern culture.

Today, all adults need to "re-genius" themselves while stopping the process of "degeniusing" our children. Modern technology and our expanded consciousness have provided the tools and information required for this monumental restructuring—both as individuals and as a society. All we need to do is wake up, make a conscious choice and step forward on the path—internally and externally—right now.

**2.6** "I live on Earth at present, and I don't know what I am. I know that I am not a category. I am not a thing—a noun. *I seem to be a verb*, an evolutionary process—an integral function of Universe."

*By Velcrow Ripper*

GUEST COMMENTATOR

We are all verbs. Not a one of us a noun. Not one a fixed identity. Thank you Bucky, for articulating something so critical, so crucial, so clearly. The liberating power of this deep understanding is a game changer.

Somewhere a few centuries back, we developed the fragmented, Newtonian worldview that haunts us still. With help from the new power of the zero, we accelerated our reduction of the world to facts and figures, cleaving it apart with a sword of Hubris. G~d died, or was killed, and in Her place we erected towers of steel and concrete. The ego made manifest, set in concrete, immobilized.

So different from the curving, integral sweep of Bucky's geodomes. Our very perception shifted, as we fragmented ourselves, specialized ourselves, as we created a binary world of black and white, of us and them, of either/or. Nature and spirit a distant other. There was security in this view. A false sense of security.

Look up! The skyscrapers are falling. Rapid climate change, economic collapse, ecological collapse, political instability, and technological escalation: the only thing we can be sure about is radical indeterminacy.

In the face of this acceleration, we have choices to make. We can freeze in fear, becoming paralyzed. Static. Resistant. Frozen. We can buy stuff or watch stuff or eat stuff, anything to avoid feeling. There's too much pain out there. We can fight like we've never fought before. If all else fails, we could always try bombing something. That usually works really well.

Or we can rise onto our surfboards and surf the power of this wave ~ particle ~ wave of change. This Tsunami of transformation. We can dive into the frothing waters with joy, celebration, and Love, following the currents, not fighting, not resisting, yet not succumbing ~ transforming poison into pearls.

Buckminster Fuller's gravestone reads, *"Call me Trimtab."* The Trimtab is the tiny rudder that trims the direction of great ships. We are not trying to shift the steel prow of the oil tanker that is industrial growth civilization, which is clearly on a collision course with limits, heading for the next spill. The next crash. The next dead end. Instead, we are becoming the collective Trimtab for Spaceship Earth. We are learning to meet the raging tides of this age of extinctions with radical grace.

We are schools of fish lost at sea, seeking to change course from the bottom of the ocean up. Slowly, then suddenly, with a committed wave of our million fins, we will steer the seemingly immobile forces of top down self-destruction back towards harmony, towards Love, towards an ever evolving universe story that is as ancient as light.

There is tremendous energy to be found in these days of quantum leaping. We are facing record-breaking weather around this trembling Earth—the hottest, the wettest, the coldest, the driest. But we are also seeing record-breaking vision arising everywhere in this season of transition. Millions of people around the world are creating a new story. We are the largest mass movement in humanity's history, and we are a verb.

Welcome to the era of resilience. Of fluidity. Of flexibility. Balanced with the strength of right relationship, of clear intention. Free will in service to the Universe made manifest on Earth. Which is Us.

The rigid, the fixed, the unmovable—they will be moved, regardless. Most likely they will snap, crackle, crumble, unable to bend with the winds of evolution. Unless they (who are Us) learn that we are truly verbs. How beautiful this understanding that we are evolving. There is so much joy in this—so much meaning. How can we not help but be in awe of the stupendous fourteen-billion-year journey that has brought Us to this place of consciousness, of conscience, of self-aware Love? A miraculous mirror reflecting the infinite journey back to the creative life force Ourself.

I am in Love with Life. I can't get enough of it. I am in Love with this pearl of a world. I am in Love with humanity. I am in Love with our wisdom—and our folly. To those who say the planet would be better off without Us, I ask that they reconsider. For we are integral to this planet. We are Earth.

We are in the midst of a great Love story, and part of that story currently involves a separation. Yes, we have lost our way. We have strayed far from our Love. But still we carry the torch, burning away, buried in our heart of hearts.

We are in a reckless mid-life crisis, spending all our resources on some useless, big red Ferrari, racing away from compassion and responsibility, lost in denial, searching for something we've always had. One day, may it be soon, we'll crash the damn thing one last time, and come back home, to our true Love, to our true Life, with a much deeper appreciation for all we have left behind. Carrying new gifts, borne of the experience of separation. And in that return, we will Love like never before.

Love is a verb. It is something we do, something we live, something we are. Every dancing cell is alive with Love. The stars are burning with Love. The Earth gives birth to Love, night and day. Death is part of Love. Sadness is part of Love.

The whole spectrum is Love, in action, in motion. Even our illusion of rigidity—born of fear—is all about Love, about our vulnerability. Our human vulnerability. We are afraid of truly living, of truly Loving, for to Love is to accept that one day, the Lover will be gone. To open the heart is to be deeply vulnerable. But to be vulnerable is to flow, to be open, to give and receive.

Yes indeed—all is impermanent, all will be lost. And that is the ultimate source of liberation. So don't hang on—but don't let go. Breathe it all in! Don't miss out by numbing down or dumbing down or running away. No matter what is happening, you are Loved, and you are Love. And it will all disintegrate, dissolve, decompose, be gone in a flash.

Don't turn from the journey! The greatest show on Earth is this very moment. This very breath. This very heart beat. This endless Love. Welcome home to Planet Earth. May you Love the Life you Live.

VELCROW RIPPER is an award-winning filmmaker, writer, and public speaker, with dozens of films under his belt, including the acclaimed *FIERCE LOVE TRILOGY* of feature documentaries, an epic exploration of the global zeitgeist during the years 2000 to 2012. It began with *SCARED SACRED*, winner of the Genie Award (Canadian Academy Award) for best feature documentary. Part two is the newly released *FIERCE LIGHT: When Spirit Meets Action*. The trilogy will conclude with *EVOLVE LOVE: Love in a Time of Climate Crisis*. His films are released widely, in theatres, broadcast, and on-line, internationally. He can be reached at transparentfilm@warpmail.net or his website, www.velcrowripper.com.

## By *Thomas Myers*

GUEST COMMENTATOR

The dynamic "verb" of a human body in motion is easily understood as an active tensegrity structure, always adjusting the forces of tension and compression so that they are perfectly balanced. When I studied World Game with Bucky in the late '60s, I learned about tensegrity and helped build a dome or two, but the general systems work warmed me more than the engineering. In the early '70s, I ran into Ida Rolf, whose able hands set me to work with the job I have held ever since: changing people's posture toward greater efficiency, length, sturdiness, and balance.

I am told that Ida Rolf met Buckminster Fuller once, but that the meeting did not go well. "*I stand up straight; I was in the Navy,*" Bucky told her, completely misunderstanding her approach to spatial positioning. Some years later, Rolfing instructor Jim Asher gave Bucky a structural integration session late at night on Bucky's hotel bed, after a lot of palaver getting him down to his skivvies. Bucky had a bad hip, but he was seen next morning in the hotel lobby, jumping up and down on it, delightedly saying, "*See, it works! See, it works!*"

Bucky always told his audiences, "*don't try to reform people, reform the environment and the people will reform themselves.*" It is odd, remembering this idea, that I should take up the work of literally trying to "re-form" the most useful tool and proximate environment we will ever have—our bodies. I have termed this attempt to change body shape (and thus increase its syntropic behavior) "Spatial Medicine"—in contrast to the currently popular Material (or chemical) Medicine of drugs, herbs, and supplements, or the Temporal Medicine of psychotherapy or shamanism.

Other attempts to reform the body itself have focused either on shifting the bones directly (chiropractic and osteopathy, which have deep roots in folk culture but were only organized into sciences in the last century) or via training the muscles (whose most ancient forms are yoga, martial arts, and folk dance, though modern personal training methods such as Pilates, Alexander Technique, and the various fitness approaches clearly belong here also).

Ida Rolf's method was uniquely different: she addressed the sinewy parts of the extra-cellular matrix of soft connective tissues, popularly known as fascia, strategically lengthening and reducing planar adhesions in this plastic medium, which in turn leads to shifts in the bony relationships and retuning of muscle tone across the body. The goal is to change our literal and figurative "attitude," deliberately conflating the biomechanical with the psychosomatic.

Changing the relative position of the "struts" through adjustment of the "tension members" cried out for tensegrity modeling, though to this day I am unsure that the human body on this macro level fulfills the strict definition of tensegrity, despite the useful modeling that is coming out of applying this metaphor (see illustration).

At the atomic level, of course, the body is a tensegrity structure, as everything material is. Dr. Donald Ingber of Children's Hospital in Boston (http://www.childrenshospital.org/research/ingber) has ably demonstrated tensegrity functioning in the inner working of our cells as well as the way that they are all suspended in the extracellular matrix via transmembranous integrin connections. This has been dramatically rendered by John Liebler at XVIVO (http://www.xvivo.net/the-inner-life-of-the-cell.)

The original and most forceful voice in favor of defining the body in terms of strict tensegrity has been the orthopedic surgeon Dr. Steven Levin (http://www.biotensegrity.com). His imagination has been ably rendered in sticks and bungee cord by designer and engineer Tom Flemons, inventor of, among other things, the tensegrity Skwish toy (www.intensiondesigns.com).

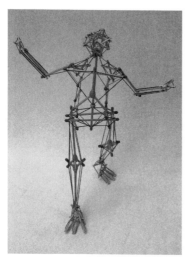

Model of human being as a tensegrity structure.
Model and photo by Tom Flemons,
www.intensiondesigns.com

Less tested but equally intriguing is the work of British os-
teopath Graem Scarr (http://www.scribd.com/doc/33438210/
Cranial-Vault-as-a-Tensegrity-Structure-by-Scarr) who has
posited that the bones of the skull are held apart by a tensegrity configuration of the membranes around the brain across
the wormian sutures.

Considering the body on a macro level—bones as struts,
organs as incompressible but deformable balloons, and connective tissue as tension members with muscles to immediately
adjust the tension and nerves to shorten and lengthen the muscles—the idea of utter and strict tensegrity is hard to assert.

To turn the leg into a tensegrity tower, Tom Flemons has
struts from the upper member coming below the upper struts
from the lower member—a requirement to support the upper
member in tension rather than compression. Examining the
actual human knee, it is hard to see that kind of mechanism
at work: the tibial plateau simply does not extend high enough
for any soft-tissue fiber to go down from there to the lower
edge of the femur. Close examination (which I have made in
the dissection lab) does not reveal any substantial fibers that

could possibly support the weight involved. The finger and toe joints present a similar problem.

The shoulder and the spine are much easier to imagine as complete tensegrities; indeed, the shoulder simply cannot be modeled in the old leverage models, and the spine is a very poor continuous compression "column." Any static tensegrity modeling will of course fall short of a living human body that continually adjusts in movement in response to psychosocial as well as biomechanical demands. Discussions of such issues can be found at http://floatingbones.com.

So if we ultimately cannot place the human body in the category of pure tensegrity, we can certainly place it firmly in the family of "tension-dependent" structures. Eliminate the soft tissues and the skeleton would clatter to the ground; it has no interlocking stability of its own. Like a sailboat or the mast-and-membrane structures of Frei Otto—which fail the pure tensegrity test, but are still tension-dependent—the body apparently makes use of a variety of engineering mechanisms in its daily rounds and extraordinary feats, from the leaf spring of the foot arches to the floating eyeball.

This imagery has very practical application to the world of rehabilitative manual and movement therapies, where chronic injury travels, and the area of binding may be far from the site of pain. Ida Rolf's "Where you think it is, it ain't" nails the body's tensegrity, which is a strain distribution system, rather than a strain focusing system. The bones float, isolated struts in a sea of continuous tension. The body's specific soft-tissue geodesics are outlined in the book, *Anatomy Trains*, which owes a good deal in both tone and substance to the expansive genius of R. Buckminster Fuller.

TOM MYERS is the author of *Anatomy Trains* (Elsevier 2009), and *Fascial Release for Structural Balance* (North Atlantic 2010), and numerous articles for trade magazines and journals. Tom directs Kinesis, which offers continuing education to manual and movement therapists worldwide. Tom lives, writes, and sails on the coast of Maine with his partner Quan and her animals, and can be found via www.AnatomyTrains.com.

⚠ IF A PERSON BELIEVES HERSELF TO BE A THING (A NOUN), she limits herself to her physical body and reality. Buckminster Fuller contemplated this possibility at great length over many years before he made his now legendary, profound proclamation, *"I seem to be a verb."*

Many people commented that Bucky was a very active verb. Even though he was a master at simply being, those who spent time with him realized that by being completely present to whatever was happening, Bucky created the space for others to have ah-ha awakening experiences.

Still, he was continually in action on a basic yet elevated level that exemplified being *"an integral function of Universe."* By demonstrating rather than telling how to be a genuine citizen of the world and Universe, Bucky once provided a viable example that experience is far more important than explanation.

We humans are not things. We are constantly changing evolving entities who are much more than our physical bodies and the "stuff" we believe that we possess.

The physical me can see, smell, taste, touch, and hear, but it can't experience emotions, take action, or come up with new ideas. The inactive noun perspective of life overlooks all the pleasure, pain, and activity that makes us human. The active verb perspective opens doors beyond what most of us can imagine.

By opening those doors for himself and others, Bucky was able to attain a perspective well beyond what most others can even imagine. Then, he brought much of that vast vision back into the realm of the noun/thing by creating artifact inventions that modeled non-physical reality.

From Bucky's perspective, all people are verbs, and today many people are very active verbs. Being busy does not, however, make one an integral function of Universe like Bucky. Being busy simply makes a person active. The real challenge is to be consciously active as Bucky was.

He was not simply going faster to get more done. He was increasingly active in harmony with Nature and Universe and in support of all life. That was, and still is, the critical component we each need to cultivate if we are to make our unique contribution to humankind and all life.

We're all doing a lot, and we're creating much more than our ancestors imagined possible. We're in the process of continually expanding our range of possibilities. We do, however, need to be aware so that we consciously evolve with and for all beings rather than selfishly do all we can to accumulate more for ourselves, our family, our nation, those who belong to our religion, or any other exclusive group.

Perhaps our new motto should be, "verbs of the Earth unite." We have moved into a period when it's everybody or nobody and being an *integral function of Universe* is no longer optional. It is mandatory for all of us to fully participate and share our talents and gifts just as Bucky did following his awakening to his "verbness."

# Purpose Without Control

## Chapter Conclusion

**2.7** **"Anyone who thinks that humans on this Earth are running Universe or that Universe was created only to amuse or displease or bore humans is obviously ignorant."**

△ JUST BECAUSE WE CAN, DOESN'T MEAN WE SHOULD _____. (Fill in a behavior you do out of habit and later regret.) Our actions have consequences, but humans are not running the show. After decades of personal study and thought, Buckminster Fuller was very clear that humans are a very tiny component of Universe. He recognized that Nature / God / Great Spirit / Higher Power / Allah / Greater Intelligence is running the show with a Grand Plan that is well beyond the scope of human understanding. He also realized that people who believe that they or any other humans have the power to fully control Universe are not fully aware.

Even such basic processes as breathing are not in our control. And we are certainly not in control of Universe, our tiny Spaceship Earth, or our environment. We can't "own" anything, and we certainly can't change anyone else's behavior. We can, however, modify our own personal behavior to be in harmony with Universe, and that is precisely what Bucky Fuller was able to do. That is also how he was able to be so successful in making a huge difference in the lives of so many people.

Society now recognizes Bucky's achievements far more fully than when he was alive. He realized that all Creation was not manifest to *"amuse or displease or bore humans,"* but we humans were manifest to serve all Creation, Universe, and the other life forms that are also manifestations of Universe. This perspective shifts the focus from me to you and the grand glory that you bring. It represents the manifestation of an awakened being who actually does unto others as he would have others do unto him.

Surely we can all gain much from looking to adopt this view more fully. It is the perspective that is most needed today as humankind continues on the path of our *"final examination"* to determine if we are a *"worthwhile to Universe invention."* We are an invention of Universe, put here to be of value and use to a Greater Glory than we can possibly imagine. And we have been given the tools to do just that.

Each of us has this latent potential, but few of us have recognized—much less used—it. Fortunately, we have people like Bucky Fuller who have forged a path of awakening and service for future generations—like those of us who currently serve as the crew on board Spaceship Earth. Now, we need to wake ourselves up and begin freely sharing our gifts just as Bucky did. That is our mandate and path as we shift from a period of competition and war to an era of cooperation and peace for all people.

## CHAPTER 3

# BEING

## SHOWING
## UP FULLY

## 3.1  Dare to be naïve.

    ⛄ BUCKMINSTER FULLER WAS CONSTANTLY REMINDING HIS audiences that he was simply an average, healthy human doing what he saw needed to be done. He was actually an ordinary human doing extraordinary things that needed to be done, and one of the primary reasons he was able to achieve so much was his naïveté and humble confidence.

Rather than coming from a rigid view of being a knowledgeable expert, he adopted an attitude of constant curiosity. Even in the last years of his life, when he was sought after as an expert in many fields, he continued to learn and grow. And that continual evolution led him to say, *"Don't try to make me consistent. I am learning all the time."*

Although many people consider consistency an important characteristic, Bucky found that consistency restricted his curiosity and focus on being naïve. Early in his life, he realized that everything was changing and so he allowed his thinking and view to change as he learned. Thus, he focused on curiosity rather than consistency. He even continued to rewrite "The Lord's Prayer" over and over as his perspective changed.

That is not to say that being consciously naïve and inconsistent interfered with his integrity. Bucky defined integrity as the ability to hold shape, and he did just that regardless of external circumstances. He continued to be a *"local information gatherer and problem solver"* on the planet he named Spaceship Earth, and that single-minded attention led him to be what we now label a "lifelong learner."

His naïve inclination did not, however, make him an easy target for people who preyed upon others because they did not yet possess his view that we live in an abundant, generous Universe. Although Bucky freely shared his knowledge and resources, he also protected his work so that future generations could learn from his mistakes and experiences.

For example, following his invention of the geodesic dome, he hired the best patent attorney he could find to write unbreakable patents for the dome. He did not do this to make money. He created unbreakable patents to protect his designs from the corporate giants that he knew would swoop in and try to bypass his patents if they saw any possibility of making money from the geodesic dome.

The result of Bucky's foresight was that when the geodesic dome was recognized as the only solution to problems faced by the US government and large corporations, Bucky charged them a great deal to use those patents. However, when groups of curious young people asked to use those same patents on their unfunded projects, Bucky never charged them.

Bucky also continued to recount the importance of naïveté and inconsistency in his writing. In 1976 at the age of 81, he wrote the epic poem "How Little I Know," sharing the fact that the older he got the more he realized how much he still did not know. This idea of knowing that one does not know is characteristic of the genuine wisdom Bucky exhibited in every aspect of his life. The following is an excerpt from that poem.

*"It is understood*
*That if you know that I know*
*How to say it 'correctly'*
*(The exact meaning of which*
*I have not yet learned)*
*Then I am entitled to say it*
*All incorrectly*
*Which once in a rare while*
*Will make you laugh.*
*And I love you so much*

*Whenever you laugh.*
*But I haven't learned yet*
*What love may be*
*But I love to love*
*And love being loved*
*And that is a whole lot*
*Of unlearnedness."*

During these most challenging times in the evolution of our species and in many people's personal growth, we can use more laughter. Daring to be naïve helps us all to recognize the unique nature of our situation and that of our entire global society. As the saying goes, "Now is the time, and we are the people."

Hopefully, each of us has learned enough to know that none of us really knows much of anything compared to the infinite possibilities of Universe. Only then will we be able to competently question authority so that we can stand up and be counted as "we the people" who are a force for good and positive change in the same way Bucky was during his life on Spaceship Earth.

### 3.2 "You have to decide whether you want to make money or to make sense because the two are mutually exclusive."

*By Randolph L. Craft*

GUEST COMMENTATOR

One of my favorite quotes in my sustainable socio-economics "slant" on Bucky's work is Bucky Fuller's "absolute law" as it is expressed in this quote. When Bucky said this at "The Future of Business" conference in 1981, he said it with a deliberate "push" in his demeanor, paused and stared at all of the audience for seemingly endless seconds, and then repeated himself—finger raised in almost a scolding gesture with more conviction than the first time. He was serious about this! I have it on videotape.

From my perspective, this is one of his most misunderstood statements—even by many of the "Bucky people" out there. To truly understand this quote, one needs to know Bucky's definition of wealth, which I have come to call "True Wealth," as distinguished from the popularly held vague definition. Bucky's definition is, *"what we have organized to take care of how many lives for how many forward days."*

First of all, we realize that the concept of money and wealth is a human concept. One doesn't experience wealth if one is dead. Wealth combines resources in the physical world with knowledge and know-how.

A tree in the forest has no "wealth/money" value if there are no humans to "utilize" it. When a human comes along and knows how to "harvest" it, the human can burn it for heat to stay warm and to cook food. The tree only then has value to the human and can be used to trade and barter with other humans for things needed to survive. If the human knows how to build a shelter with the tree—additional knowledge—the tree then takes on additional value. The tree didn't change, the human knowledge changed, and the "wealth" value increased.

Money, on the other hand, is a tool humans invented to represent "wealth." Money is a contract between two or more people who agree that the "tool" has a certain value. That value is maintained only as long as all parties trust in that value.

Money isn't wealth. Money only represents wealth. Money is basically trust.

As time went on following the invention of the concept of money, humans forgot about the connection between money and the wealth it represents, and we started thinking that the money was the wealth. Once one thinks money is wealth, your whole thought process shifts. You will focus on "making money" instead of "making sense" by producing the "true wealth" that the money represents.

Producing "true wealth" makes sense—in any economy anywhere under any conditions. "Making money," on the other hand, usually has people coming up with all kinds of schemes and manipulations that usually make little to no sense when observed from the perspective of "true wealth."

The US/Western economies are currently a total demonstration of the "making money" thought process, and the "emerging" economies of Spaceship Earth are not far behind. "Get rich quick at any cost" (win/lose or lose/lose), making little to no sense, has replaced a deliberate and thoughtful process to produce needed value for others at a fair value-added profit. This sensible way of operating endures, can only grow, and makes total sense. Thus, Bucky's statement *"you have to decide whether you want to make money or make sense because they are mutually exclusive"* helps us all to evaluate our actions and move toward creating a more sustainable society.

RANDOLPH L. CRAFT has been studying Buckminster Fuller's work for decades and teaches about "true wealth" around the world. To learn more about him, "Sustainable Global Socio-Economics" and "True Wealth," based upon the life and work of Dr. R. Buckminster Fuller visit www.FullerEducation.org.

△ FREQUENTLY REPEATED AND GENERALLY NOT UNDERSTOOD, this quote represents a critical aspect of Buckminster Fuller's operating strategy and the way he showed up after his 1927 contemplation of suicide. At that time, he had done his best to make money and support his family, but he was a total failure in the eyes of society.

On the verge of killing himself, he had an epiphany/spiritual experience. Not knowing what to do next, he entered into several months of silent meditation, contemplation and pondering his purpose. The result of that period was his great experiment to determine and document what a single individual could accomplish that could not be achieved by any corporation, government, or institution no matter how large or powerful. One of the key elements of that "56 Year Experiment" in which he used himself as "Guinea Pig B," was his resolve to never again work for a living.

Bucky decided to make sense rather than to make money. He did not, however, decide to starve, be homeless, or shirk his responsibility to support his family. Instead, he resolved to keep his intention focused on making sense for the benefit of all people rather than trying to make money for himself and his family at the expense of others as most people were (and still are) doing.

His discipline could be restated as follows: When embarking on any new endeavor, a person whose intention is to make money will probably fail while a person whose primary intention is to make sense and contribute to others will most likely succeed and be supported financially. It's not about the overall action or money. It's really about the intention, especially the initial intention. And this is what drives the entire endeavor.

Over the course of his "56 Year Experiment," Bucky proved that this strategy works well. In the 1950s he had an annual income of over one million dollars. Realizing that Nature never hoards and Nature only saves enough to sustain a project,

Bucky quickly "reinvested" his income in the next not addressed project he saw needed to be done.

Bucky's rationale for trying this strategy was quite simple. As humans, we are all part of the great whole we call Nature. Nature always supports whatever is supposed to be. There is no artificial in Nature. Nature always and only allocates the exact amount of resources necessary for any "project," be it a tree, a rainstorm, or a human.

Nature never uses too few or too many resources. If something is not in sync with Nature, then is not supported. In fact, it does not even exist.

Humans are a unique species in that we often attempt to create projects that are not in sync with Nature. These always fail. They may appear to be working for a while, but anything that is not in accord with Nature and her generalized principles will fail.

This is particularly true of projects or inventions that are initially created to make money and not make sense. Consider the internal combustion engine. In its short tenure (a century is a spit in the ocean when considering the great cosmic time scheme), the internal combustion engine and the fossil fuel it requires have destroyed a great deal of our planet's natural resources and habitat. That invention is literally killing us and much of the life on Earth.

Nature and Earth will survive with or without humans. Because it was created to make money rather than make sense, the internal combustion engine does not support the success of life on Earth and will fail. And it may also result in the extinction of the species that created it.

Bucky realized that Nature is always successful in every aspect of her work, and he modeled his actions on the thriving templates he found in Nature. At this critical juncture in humankind's evolution, when we stand on the verge of extinction as a species, and in an era of what Bucky called *"humanity's final exam,*" "we the people" must always make sense rather than trying to make money at the expense of other people.

This is not some pie-in-the-sky idea. It's the mandate of a man who devoted the last 56 years of his life to finding out what works and how we can all implement such best practices into our work and our lives.

**3.3** "So I vowed to keep myself alive, but only if I would never use me again for just me— each one of us is born of two, and we really belong to each other. I vowed to do my own thinking instead of trying to accommodate everyone else's opinion, credos, and theories. I vowed to apply my own inventory of experiences to the solving of problems that affect everyone aboard planet Earth."

*By Dr. Joel Levey*
GUEST COMMENTATOR

During a time of tremendous personal crisis in his life, Buckminster Fuller nearly resorted to taking his own life. Instead, a very profound and transformative experience awakened him and inspired him to rededicate his life to being of benefit to others. When writing and speaking about this experience, he made the above statement.

In affirming that "we really belong to each other," Bucky echoes an insight from Brother David Steindle-Rast, who says, "Ethics is how we behave when we realize that we belong together."

Bucky's heartfelt dedication reflects the universal principle that the Buddha called "conceiving the mind of enlightenment." This quality of being is the core motivation of a *bodhisattva*, a person who dedicates him or her self to realizing their true nature and highest potentials in order to support all beings to do the same.

As we awaken from experiencing ourselves as separate from others, our wisdom deepens to behold a view of the profound interdependence of all beings. From that view, it is natural to feel empathy for others and to want their happiness. You would never intentionally harm another or take advantage of them for your personal benefit but instead would wish for them to be safe, free from suffering, and to realize their highest potentials.

As Martin Luther King Jr. once eloquently stated, "All life is interrelated... caught in an inescapable network of mutuality tied in a single garment of destiny. Whatever affects one directly affects all indirectly. For some strange reason, I can never be what I ought to be until you are what you ought to be. You can never be what you ought to be until I am what I ought to be. This is the interrelated structure of reality."

The great Indian sage Shantideva expressed this realization as, "How wonderful it would be when all beings experience each other as limbs on the one body of life."

If we were to emulate Bucky's vow in our own lives, we would dedicate ourselves to:

- a life of selfless service;
- a life of mindfulness and vigilance regarding what our motivations are moment to moment throughout each day, lest we lapse into old, self-centered habits; and
- a life of ceaseless devotion to searching for solutions to problems that plague those who share the life-support systems of Spaceship Earth.

Keep in mind that although it may be impossible at this stage of your evolution to live perfectly true to such high ideals, holding such an aspiration as the core organizing principle of your life provides you with a frame of reference/reverence to keep returning to whenever you lapse into the habit of living for just yourself. If you are clear on the direction or goals you are dedicated to in your life, you'll be more likely to notice when you drift off course, and keep returning again and again to align and attune to the path you are dedicated to following in your life.

If you find this a compelling possibility for your own life, I invite you to take a step in this direction today. Here are some simple ways to begin:

- Experience the next breath in a way that is dedicated to all beings remembering their intimate connection with the web of life. As you inhale, envision that you and all beings receive what you most need in this moment to flourish in your life, and as you exhale, dedicate that letting go to supporting all beings in releasing their tensions, misconceptions, self-centeredness, ad infinitum.
- Another simple step would be to mindfully drink your tea, or sip your java with the motivation that, "May this liquid satisfy the deepest thirst of myself and all beings."
- As you open a door to step outside, envision that you are leading all beings to greater freedom and openness.
- As you return home, envision all beings having a place of peace, safety, and refuge to abide within.

As you practice in these kinds of ways, you will gradually develop your capacity to live ever more fully so that your every thought, word, and action is dedicated to benefiting others. Though at first this may be largely a discipline of aspiration and imagination, over time, you will find that you have transformed your identity, worldview, values, and way of life to truly embody Bucky's vow to live for the benefit of all.

As we reflect upon Bucky's brilliance and aspire to walk in his footsteps, may we hold Bucky's hand, sense the light of his inspiration filling us, and align our hearts and minds with his inspiring example for the benefit of all!

JOEL LEVEY, Ph.D., is co-founder of WisdomAtWork.com, International Institute for Mindfulness, Meditation, and MindBody Medicine, and the International Institute for Corporate Culture and Organizational Health. With his wife, Michelle, he stewards the

Kohala Sanctuary, a permaculture-designed learning center on the Big Island of Hawaii, and is the co-author of many books and audio programs including: *Living in Balance*, *Luminous Mind*, and *Wisdom at Work*. Joel's work with leading organizations around Spaceship Earth draws insights from science, spirituality, and medicine, and offers an integrative approach to change resilience, and personal, community, and organizational sustainability and thrivability.

⚠ MOST OF US BELIEVE THAT WE ARE INDEPENDENT AND think for ourselves, but we're generally under the spell and control of many external forces. Although when asked we often say that we're doing our best to be in integrity, we usually don't have the integrity of holding our shape (knowing and maintaining our core values) regardless of external factors and circumstances. We're swayed by the opinions of others as well as the flood of information that inundates us from morning until night.

In this 1927 declaration of his integrity and mission, Buckminster Fuller vowed to do his best to maintain his integrity, knowing that his radical stance would be challenged again and again. He was making this vow at a time when people were far less conscious of themselves and their environment than is true today. They were also far more isolated from the majority of the people on what we now know is a tiny Spaceship Earth.

So, to proclaim that he was going to *"apply my own inventory of experiences to the solving of problems that affect everyone aboard planet Earth,"* was, and still is, an immense commitment. To do this publicly, as Bucky did, ramps up the intensity even more, yet this is the calling that is required of us at this critical moment in the final examination of humans as a species.

Many of us are now aware that we're all in this together. There is no other planet that we can escape to. Once the air or water on Earth is polluted beyond the point of being supportive of human life, there is no turning back. And if we reach a point where everybody believes the ideas and vision thrown down to us from above by corporations, governments, and other organizations that would have us live in fear, believing that there is not enough and we must compete with one another, the game of life on Spaceship Earth is over.

Bucky's commitment rings true today more than ever. We each need to trust our inner judgment and personal experience above all else. The data is available to each of us with a few clicks of a mouse, and we need to examine it and make our own decisions. Then "we the people" need to rise up and unite in a single voice calling for unity, peace, freedom, justice, and abundance for all. This is the vision of "a world that works for everyone" that Bucky championed throughout his adult life, and this is the only solution that is now acceptable if we are to survive and thrive.

We really do "belong to each other." It is time for each of us to be our sister's and brother's keeper. It is time to stop the needless death of fifty thousand people every day because they have no food to eat on a planet that produces enough food to feed everyone. It's time to cry out STOP to those who generate great wealth for themselves by continuing to promote war and destruction.

"We the people" can take charge of our planet and ourselves. We can think for ourselves, and we can go beyond the opinions, credos, theories, and beliefs that have led us to this brink of destruction. Fortunately, we have the lives of people like Buckminster Fuller to use as a template for our actions. We most certainly don't want to blindly follow in his footsteps, but we must examine his great work so that we can think for ourselves and decide what we can learn and apply from his "56 Year Experiment" on behalf of all people.

 **3.4** **"Never mind if people don't understand you, so long as no one misunderstands you."**

*James Roswell Quinn*

GUEST COMMENTATOR

In 1984, a good friend gave me a copy of Buckminster Fuller's book *Critical Path*. At first, his words nourished my hunger to make a greater impact with my life. But soon, I realized this amazing man had created an instruction manual for how to create positive change ... locally and globally. I have been a student of Buckminster Fuller, and a teacher of his concepts, ever since.

People who have attended my personal growth and leadership seminars will likely be surprised that I have chosen this quote as my contribution to this important book. Most would probably have assumed that I would speak on Bucky's concepts of "Precession," "Synergism," or the "Trim-Tab" principle (which forms the introduction to my book *The Love-Based Leader*). While these and many other of Fuller's concepts are dear to my heart and important in my teachings, this one quote seems to go to the heart of solving difficult personal and global problems.

To me, no matter how hard we try, it has always seemed that the normal result of most attempts at communication is misunderstanding. In this event, resentment and justified reactions become commonplace. In fact, misunderstanding is at the root of all human conflicts, from marital discords to international wars.

For example, imagine that I have just asked you to pick me up at the airport at 6:00 PM. If you understood me correctly, you will be there on time. No problem.

If you did not understand me, you would stop me and ask me to repeat it. Again, no problem.

But, if you misunderstand me and think I said to pick me up at 7:00 PM, then you would not ask for clarification. You would think you got it right. As a result, there will be hurt feelings or worse when you fail to show up on time. There will be a problem.

Bucky realized this and strove to be so clear in his own communications that he had to speak in a new way. Fuller actually had to invent words and use creative grammar in order to state his ideas in a way that gave people a chance to understand his concepts. Often called "Fullereze," his unique language was difficult for many to understand.

I do not think this bothered Bucky because, in the end, very few people "misunderstood" him when he explained something. People either understood what he was telling them or they did not. But nobody ever went away with the wrong idea.

Some understood his concepts immediately. They began to question how to implement what they learned and wanted to learn more.

Others did not understand his concepts at first. They usually became confused in a manner that demanded answers. They then asked clarifying questions or studied further, which ultimately led them to understand.

The question becomes, "Do you want to make a bigger difference with your life?" The answer is for you to strive to eliminate miscommunications.

When you tell someone something significant, such as a concept you are striving to convey or a personal request, get into the habit of asking clarifying questions. Even when it seems like they understood, ask them to tell you what they heard. Make sure.

Conversely, when someone else is striving to communicate an idea or a need to you, get into the habit of repeating back to them what you thought they said. Make sure.

It is with absolute clarity that we can create trust between individuals and between nations. What kind of a world do you want to live in?

Therefore, if you want to be a part of creating (in Bucky's words), *"a world that works for everyone,"* then strive for clarity in everything you say and everything you hear. Become accountable for your results and teach others to do the same. The world has enough victims; we need leaders.

As my wife often says, "Clarity at all costs, or it costs you all."

I think Bucky would agree.

JAMES ROSWELL QUINN is an author, keynote speaker, leadership trainer, success coach, and financial educator. Since 1979, he has made over fifteen hundred presentations to tens of thousands of people in ten countries. The body of Quinn's work includes his book, *The Love-Based Leader*, and his eight-CD personal audio seminar, *Get Over Yourself*. He has co-authored the book *Speaking Of Success*, with Jack Canfield, Ken Blanchard, and Steven R. Covey. More information about his works is available on his website www.JamesRQuinn.com. He can be contacted at GlobalKeynote@aol.com 815-248-2081.

△ THIS DISTINCTION BETWEEN NOT BEING UNDERSTOOD AND being misunderstood was a critical aspect of Bucky's personal discipline and operating strategy. After many agonizing experiences of being misunderstood, he came to realize that he was much better off if people simply did not understand him rather than being misunderstood.

Thus, he would go to great lengths to clarify each and every detail in his explanation of almost everything. There is no "short and sweet" answer in Bucky Fuller's world because everything needs to be clarified so that it is not misunderstood.

This is apparent in his lectures as well as his writing. The lectures were actually labeled "thinking out loud" sessions, and they often ran on for hours as Bucky clarified each point he made. The result of those lengthy discourses in both speaking and writing was usually people not understanding, and that was acceptable because Bucky discovered that not being understood would usually result in people thinking about his ideas even more and asking clarifying questions.

Bucky also knew that if he covered topics in a cursory manner, people might misunderstand his concepts and take inappropriate action or share incorrect insights based on their misunderstanding. So, Bucky's writings and talks are unusually convoluted, lengthy, and sprinkled with words he created when he could not find a word that expressed the specific reality or concept he was explaining.

One such example is the words "sunsite" and "sunclipse." Bucky coined these words to challenge what he often said was the greatest lie we were foisting upon our children— that the Sun rose and set. The Sun does not rise or set, and just saying or reading those words creates the erroneous belief that humans are the center of all reality around which everything revolves.

This is not a view of reality that Bucky wanted imposed on young people, so he created the word "sunsite" to replace sunrise and "sunclipse" to replace sunset. These new terms allowed him to correctly speak of the phenomenon of the Sun's becoming visible and not visible to humans as our Earth rotates.

When people initially encounter these words, they might not understand them, but that usually stimulates their curiosity and causes them to ask questions. Thus, the truth emerges from not understanding rather than being misunderstood.

**3.5** **"There is something patently insane about all the typewriters sleeping with all the beautiful plumbing in the beautiful office buildings—and all the people sleeping in the slums."**

△ THE ISSUE OF RESOURCE ALLOCATION WAS A KEY ELEMENT in Buckminster Fuller's thinking and action, and he was continually pointing out irrational cultural activities in this area. More than just a critic or social commentator, Bucky followed his own advice of doing what needs to be done and is not being attended to by taking action to solve the problems he witnessed.

His initial action was generally to inform and educate as he did when he talked about office buildings being empty while people slept on the streets. Next he created artifacts (inventions or other physical objects) to solve the problem, mirror Nature's generalized principles and demonstrate how Nature's principles can be applied to an issue, offering unique new solutions for the future.

Bucky's solution for housing people efficiently was the Dymaxion House. This unique round, inexpensive, self-sustaining house was preceded by the Dymaxion Dwelling Machine, which was another round dwelling made from circular grain silos.

In 1928 when he came up with his initial concept of the Dymaxion House, Bucky was financially destitute, but he was able to build a model. He then began showing that simple model to anyone who would listen in an effort to make his solution as known as possible. And, after decades of patiently waiting, Bucky's strategy of "emergence by emergency" eventually paid off in 1944 when the U.S. government was looking for ways to keep wartime aircraft factories busy.

It was then that the assembly line–produced Dymaxion House was discovered as the easiest solution available, and Bucky quickly began building a working prototype. Even though the Dymaxion House was never a commercial success, that prototype is now a permanent installation at the Henry Ford Museum in Michigan.

The reality of empty office buildings while people were homeless made no sense in 1928, and it remains just as illogical and inhumane today as people are losing their homes to foreclosure. Efficiency and inclusion dictates that we make use of all our resources as effectively as possible, and our lack of thinking outside the box and compassion is evident in this instance.

On weekends and evenings, lights and heating are often left running in office buildings and other commercial structures while homeless people are exposed to the elements on the same streets as the warm, dry buildings. In most cities, this is particularly evident in the downtown core areas, which contain the most homeless people and a majority of commercial buildings vacant on weekends and evenings.

This is just one instance of our wasting valuable resources while treating underprivileged citizens in an inhumane manner. At this critical moment in humanity's evolution, we need to maximize all our resources while minimizing waste. This would be very simple in the instance of homelessness and office buildings.

During extremely cold periods, this is exactly what happens. When a blizzard makes outdoor temperatures unbearable, public buildings are opened for the homeless, but our leaders are unwilling to take that same action the rest of the year. And so our office machinery is cared for better than our fellow crewmembers on board Spaceship Earth.

**3.6** **"Unconscious decisions have consequences. Our assumptions drive our priorities, and in many cases we don't even acknowledge they are there. Innovation arises from questioning the old assumptions."**

*By Stephan A. Schwartz*

GUEST COMMENTATOR

In 1969, I was in Washington, D.C., a twenty-seven-year-old editor of *Seapower*, a journal of maritime and naval affairs. Late one afternoon in the spring I was approached by the director of the Navy League, a non-profit foundation that sponsored the magazine, who said to me, "This is something that might interest you, Stephan. How would you like to spend the day with Buckminster Fuller?"

Fuller was scheduled to give the keynote address at the Navy League's Annual Conference. He was to be picked up at his hotel about 10 a.m. and needed to be staffed for the day. If I volunteered I was to take him wherever he wanted to go and make sure things went smoothly. I agreed immediately and was struck by the relief in the director's face. I realized he didn't want to be burdened with Fuller for a whole day, just before the conference, and was glad I would do it.

On my side I couldn't believe my luck. I had been interested in Buckminster Fuller for years, dating back to when I saw one of his early domes and a picture of the Dymaxion car. The idea of spending a day alone with him I saw not as a burden but as a wonderful unexpected gift.

Two days later I am standing in Fuller's hotel lobby when he gets off the elevator. I recognize him immediately and go up and introduce myself. He has a low flattop haircut and is wearing a bluish green spectacularly unfashionable suit of some hard finish cloth. He looks almost iconically like an inventor engineer. In fact I realize he defines the type in the way Edison must have defined it for his generation.

Fuller is smaller in stature than I had thought. His glasses are broken and he has taped them back together. He has hearing aids. I am surprised at how near-sighted he is. It doesn't show up in many of the pictures. His astigmatism is clearly much worse than my own. It distorts his eyes slightly but this does nothing to filter out a kind of intense curiosity he projects whenever he focuses on me—or anything else.

Despite the differences in our age and status, there is no awkwardness between us. He is available and engaged, asking questions and commenting. After speaking with him for a few moments I have the sense he is incredibly competent and understand that he looks at the world in a way unfamiliar to me.

I ask him what he wants to do. "I'd like to go to the Mall, have lunch somewhere you pick, and then come back to the hotel. I speak this evening."

We catch a taxi and he asks me to have it take us to the science museum. As we walk the Mall he takes in everything and begins to talk about the people, the buildings, the placement of the buildings, how the people use the buildings. "*All those stairs*," he says pointing at one of the grand Mall buildings. "*This building was designed to impress, not to be helpful or even functional.*"

Over and over he introduces me to new ways of looking at familiar scenes. It is a mix of science, engineering, and a finer higher element. I realize most of the written accounts I have read about him miss this subtler realm. As we walk, I understand I would have missed it as well except for my own experiences as a meditator. His words are the expression of his metaphysic. He sees pattern and system where most people see only isolated elements.

He has taken the time to think through the details, the design choices of our culture, to the point where he can see the higher over-arching narrative of who we are. And it is reflected in almost everything he says. He extrapolates from the present reality to what might be through practical details.

"*Do you see the Mellon Gallery?*" he asks, using the old name for the National Art Gallery, begun from a bequest from Andrew Mellon.

"Yes."

"*What does it weigh?*"

"I have no idea," I reply. It takes me a beat to even consider the question. It seems so odd.

"*Nor did the architects that designed it.*"

"Does it matter?"

He looks at me for a moment, shading his eyes with his hand. "*A marine architect would know. He would know to the pound. And he wouldn't use a pound more than was required. The sea compels such attention.*"

It has never occurred to me to consider the weight of a building, and while I think it is an interesting way to look at buildings, I know I am still not getting his point. He lets me think about this as we walk on. After a beat he picks up his thread.

"*The design decision not to consider the building's weight does not exist in isolation; it is a way of looking at the world. From that assumption of what matters and what doesn't an entire chain of consequences is set in motion. Each step of which because of the flaw in the original conception—not considering the weight—needlessly consumes natural resources and the time and energy of people. Unconscious decisions have consequences.*"

I suddenly understand his point. Almost visually I can see in my mind's eye the network of consequences. How not caring about weight leads to design choices, which lead to orders, which lead to manufacturing, which leads to fluctuations in raw material prices. I see that when unchallenged assumptions are repeated and accepted as the normal way of thinking about a problem then culture is shaped to that bias.

*"Our assumptions drive our priorities, Stephan. And in many cases we don't even acknowledge they are there. Like asking how much a building weighs, and what that means. Innovation arises from questioning the old assumptions."*

Standing there under the trees that line the Mall's length, shaded from the mid-day Sun, I know Buckminster Fuller has given me the gift of a teaching moment. And four decades later I see how often I have used his insight, and how much I owe to saying "Yes," when I was asked, "How would you like to spend the day with Buckminster Fuller?"

STEPHAN A. SCHWARTZ is a Senior Samueli Fellow at the Samueli Institute, editor of Schwartzreport.net, and columnist for the journal *Explore*. He is the author of nearly seventy peer-reviewed scientific papers and technical reports, numerous magazine and newspaper articles, producer of twenty documentaries, and the author of four books, *The Secret Vaults of Time*, *The Alexandria Project*, *Mind Rover* and, his latest, *Opening to the Infinite*. To learn more about Stephan and his work visit his websites www .stephanaschwartz.com, www.schwartzreport.net, explorejournal.com/content/schwartz.

BUCKMINSTER FULLER MADE A PRACTICE OF QUESTIONING almost everything. This was especially true of old assumptions, whether they were his personal ones or society's. Another of his disciplines was to be as certain and conscious as possible about an issue before making any decision. He did not wait until he was absolutely sure of something before acting, but he made the best decision he could, based on his current knowledge and experience, knowing full well that everything is in constant motion and changing. In other words, he followed the "ready-fire-aim" philosophy of living. Within this context, a person takes action (fire) and course corrects (aim) after each action rather than trying to find the "perfect moment and action."

Bucky often reminded people *"don't try to make me consistent. I am learning all the time,"* and he was comfortable challenging assumptions and course correcting his initiatives as they progressed. He also realized that the law of cause and effect (also labeled karma) is a vital force influencing every activity and thought. Every action (be it consciously considered or not) has a consequence for the person taking the action and others. Bucky found this to be most important for humankind as we teeter on the verge of what he labeled our *"final examination"* to determine *"whether humans are a worthwhile to Universe experiment."*

Our ancestors have made some severe unconscious decisions, and we are now facing the consequences of those choices. Most of those poor decisions were based on the best information available.

Still, these decisions were made with the best of intentions from a series of untrue assumptions including that:
- We live in a Universe where resources are scarce.
- Making and hoarding (saving) money is good.
- There are an unlimited amount of physical resources on our planet so we can use them up and throw our "waste" into landfills.
- Everything revolves around us as individuals and as a species.
- We must fight for our share of the pie.

These are just some of the false beliefs that have helped move us to the brink of a planetary environmental disaster (both globally and individually). We now know that none of these statements is true, and a growing number of us have experienced the abundance and natural flow of a Universe that supports all life. Bucky often had that experience of abundant flow and support, and he was always freely sharing the information he had learned. He often reminded his audiences,

*"It is now highly feasible to take care of everybody on Earth at a higher standard of living than any have ever known. It no longer has to be you or me. Selfishness is unnecessary. War is obsolete. It is a matter of converting the high technology from weaponry to livingry."*

Many people now strive to make more mindful decisions that fulfill the philosophy expressed by our wiser ancestors and many indigenous cultures when they talk about considering the effects of their actions seven generations into the future. "We the people" are moving toward a time when our priorities are not driven by our unconscious assumptions but by the truth of our individual and collective experiences. We now have the technology to compile and disseminate legitimate data that is authenticated by judgment-free computers. And we no longer have to trust the propaganda espoused by leaders, be they elected officials (beholden to the donors who finance their election), corporate officers, religious officials, or any other of the people who seek to gain advantage for themselves and their constituents, shareholders, congregations, or other splintered factions at the expense of the whole of life.

Our world is extremely different than it was just a few years ago. It would not be recognized by the people who did the best that they could but made some unconscious decisions that have resulted in extreme suffering for many and the extinction of numerous species on Planet Earth.

It's time for all of us to stand up and question old assumptions that no longer serve the whole and the creation of a sustainable planet that supports us all. It's time to examine our personal assumptions and release those that no longer serve us in being fully functioning global citizens. And it's time that we begin to collectively and individually make decisions that support a *"world that works for everyone"* seven generations into the future and beyond.

**3.7** "We are called to be architects of the future, not its victims. The challenge is to make the world work for 100% of humanity in the shortest possible time, with spontaneous cooperation, and without ecological damage or disadvantage of anyone."

*By DC Cordova*

GUEST COMMENTATOR

When I was twenty-seven years old, I first heard Bucky speak. Though he didn't say these specific words at the time, his life was dedicated to answering that question—and creating artifacts that would support that outcome. That kind of thinking was so radical at the time and still is for many globally! I was raised to believe that the reason we had starvation on the Planet was "God's way of handling over-population"—an idea that even as a young girl growing up in Chile, I found horrifying, and it never made sense.

When I was exposed to Bucky's teachings ...
- the way that he shared about generalized principles;
- the way that he taught us how to look/experience nature in a new way by using accurate language such as "sunclipse" as opposed to sunset and "sunsite" as opposed to sunrise;
- the way he described the mind and human behavior and about Spaceship Earth
  ... it was literally a mind-blowing experience.

My thinking and life changed radically. His teachings made so much sense. There was a part of me that understood it so well that I ended up spending my life committed to passing on his work to as many people as possible through experiential trainings—focused on business, no less.

It was out of asking the question in this quotation that my purpose evolved: to uplift humanity's consciousness through business.

And because he inspired me so much to question the status quo, my mission also evolved: to transform educational systems around the world to eradicate poverty and hunger.

Test it for yourself. Keep asking that question. See what your "phantom captain" as Bucky called the Spirit, begins to show you.

By far, this is one of the most powerfully thought-provoking ideas. Remember that the quality of your life is dependent on the quality of your questions...

I hope it works for you as well as it did for me!

DC CORDOVA is CEO of Excellerated Business Schools®/Money & You® a global organization that has over eighty thousand graduates from the Asia Pacific and North American regions in English and Chinese. She is the author of the comprehensive systems manual, *Money-Making Systems*. For more about DC check out her websites www .Excellerated.com and www.MoneyandYou.com. She can be reached at Info@excelle-rated.com or (619) 224-8880.

"ARCHITECTS OF THE FUTURE UNITE" COULD BE THE RALLYING cry of all those who seek to create Bucky's vision of a "*world that works for everyone.*" Many people believe Bucky Fuller to have been an architect, but the label he most often used in referring to himself was "comprehensive, anticipatory design scientist"—often shortened to the term "comprehensivist."

We are all comprehensivists. At least we begin life as comprehensivists, interested in everything. Just watch a toddler exploring. She will use all her faculties to investigate everything. She doesn't just look at the cat or the feather. She wants to touch, smell, hear, and even taste them. Her world is vast and open for discovery, but it quickly closes down as adults with the best of intentions attempt to "socialize" her.

Still, it takes years for our comprehensive nature to be "uneducated" out of us. We're taught and told that we must specialize in order get a good job, make a lot of money and "succeed." However, as we become more specialized, we lose the global perspective so needed at this time of intense change and crisis for humans.

A comprehensive view such as Bucky's is critical to our survival and success as individuals and as a species. We humans have used our superior minds to become stewards of Spaceship Earth, and we now need to employ our hearts in making sure that all sentient beings are cared for. We need to recognize that we can all live happy, fulfilled lives without damaging our environment or any sentient being. We need to do this despite the false pressures of fear propagated by those in power. That is the mandate for "architects of the future."

This is not some pie-in-the-sky dream. Bucky proved that we now have enough resources to care for everyone if we shift from weaponry to livingry. It's just that simple. We need to stop trying to kill each other and start making sure that everyone has food, a good education, shelter, health care, etc.

This can only be done through cooperation. We need to shift from that male-oriented culture of competition to a balanced culture where the feminine is equally respected in the management of Spaceship Earth. Then we will have genuine peace and prosperity. Bucky proved beyond a shadow of a doubt that this is possible when in 1934 he compiled the data showing that in 1976 we would reach a tipping point of having enough resources to support and sustain all life on Earth. Now, it's up to each of us to do our part in implementing success both globally and locally before we pass the point of no return and can no longer undo the harm that we have done to our planet and all its inhabitants.

**3.8** "I have learned that it is possible to stand and think out loud from the advantage of our most effective possible preparation, which is all recorded, and on tap in our brains and minds. Advance thought about our discourse spoils it. There, awaiting its anytime employment by our brain-scanning mind, is the ever recorded and highlighted inventory of our life-long experiences integrated with all the relevant experiences others have communicated to us. Out of this inventory, your live presence catalyzes my freshly reconsidering thoughts relevant to our mutual interests."

BUCKMINSTER FULLER WAS A MASTER OF SPONTANEITY. HIS "thinking out loud" presentations would often run nonstop for several hours or until the building closed for the night. He also felt that anyone is capable of communicating in a similar manner. This is done by releasing preconceived ideas and speaking from the heart in the present moment while addressing issues from the perspective of his famous *"world that works for everyone."*

To do this, however, requires another skill Bucky cultivated and utilized—the ability to be silent and listen to the relevant experiences others communicated to him as well as to his inner guidance in whatever form it arose. In other words, he was able to be completely silent and present for another person and also quiet his mind enough to recognize his inner guidance.

Although most people viewed him as an effusive speaker, Bucky could sit quietly listening to others' thoughts and ideas for long periods. This was particularly true of children, whom he recognized as having access to great wisdom because they were not yet (as he labeled it) "degeniused."

This ability to be present for another without mentally planning one's next thought seems to be a skill that is becoming less and less prominent or valued in modern society. Many people today seem to have an opinion about almost everything, and because of modern technology, they feel that they are entitled and obliged to share their opinion with everyone all of the time. Because we can now do that so easily online, when people gather face to face they embody that context of "I have something important to tell you, and I need to be heard right now!"

This is probably not true. We can all usually wait our turn to share while listening intently to another. When a person follows Bucky's way of being, she is open and interested in the thoughts and ideas of another. And from that context of curiosity, Bucky was able to have *freshly reconsidered thoughts relevant to our mutual interests*." In other words, by understanding and appreciating another person's perspective with an open mind and heart, he was sometimes able to shift his view. Then and only then would he respond to the thoughts and ideas that had been offered, and his fresh openhearted reply was often the basis of the wisdom that so many people came to hear from him and continue to read in his writing.

In addition to cultivating his ability to listen, Bucky was also a master of speaking spontaneously, and he "transmitted" both a content and context that changed people's lives forever. Like most people he had experienced a great deal, and he disciplined himself so that he could allow his next thoughts to spontaneously formulate from all his experiences—including the comments of people who had just spoken.

We each carry important "pieces of the global success puzzle" to contribute, and those gifts are usually most easily accessed in live meetings where people leave their personal agenda at the door. We all have access to solutions and new ideas in our "*ever recorded and highlighted inventory of our life-long experiences integrated with all the relevant experiences others have communicated to us.*" In other words, our brilliant minds have absorbed everything we have experienced, and any of us can draw on that inventory at will—if we simply relax and speak freely when it is our turn.

Rather than come prepared with your internal or external Power Point presentation, consider showing up to any gathering "unprepared" with an open heart and an open mind. Bucky trained himself to do just that, and he was very successful at "being Bucky" in front of thousands of people hundreds of times each year for the majority of his adult life.

The trick to such a vulnerable spontaneity lies in the context of your presentation. The intention has to be as inclusive as you can make it. Bucky was a champion of all people, all life, and our fragile planet, which he labeled Spaceship Earth. When he spoke to people, his intention was to support them in waking up to their true potential. He went to great lengths to make sure that he always maintained that perspective and never attempted to take advantage of others or speak when no one was interested or listening.

# COMING TOGETHER

## CHAPTER CONCLUSION

**3.9** **"If two of us meet and you take
a paper out of your pocket and
start reading a speech, I will
say, 'Let me have that. I can
read it myself more effectively.' I
am confident that live meetings
catalyze swift awareness of
the particular experiences of
mutual interest regarding which
our thoughts are spontaneously
formulated. Live meetings often
become pivotal in our lives."**

△ BUCKMINSTER FULLER WAS A MASTER OF "LIVE MEETINGS," and we can all learn from his dynamic, ubiquitous way of being with others. When he walked into a room or stepped onto a stage, people felt the palpable force of his presence. That force was not, however, the mindless zeal of a motivational speaker or the staunch righteousness of a "spiritual leader."

Instead, Bucky radiated the secure confidence of an *"average healthy man"* who had regained his youthful innocence and wonder by trial-and-error experimentation and following his heart path to support all life. He knew himself to be a spiritual being having a human experience, but he did not flaunt a superior position. He advised people not to believe a word of what he (or anyone else) said. He also reminded us that we have an inner wisdom and the ability to check things out against our own personal experience rather than blindly accepting the mandates of those who call themselves "leaders," "wise," or "evolved."

During his entire adult life, this message remained constant. Bucky reminded audiences that such people were most likely not acting in the best interest of all "leaders," especially if they were telling us how great they were for doing just that. In a manner similar to John Kennedy during his inaugural speech, Bucky suggested that we ask not what your Earth can do for you, ask what you can do for your Earth and all her inhabitants.

Again and again, he championed and demonstrated the power and significance of the *"average, little man."* He also reminded people that despite the enormous evolution of technology, no innovations could outstrip the potential of the individual human being, especially when we come together. Thus he firmly believed that *"live meetings catalyze swift awarenesses."* In other words, because we are functioning as physical beings on Earth, coming together face-to-face and heart-to-heart and mind-to-mind creates synergistic insights and experiences that can only happen within that context.

Very wise leaders and teachers have always known and used this simple fact. Such meetings "when two or more come together" produce a unique experience, and Bucky was quick to notice that exceptional yet somewhat elusive nature of our live gatherings. This is even more significant today when so many of us are "meeting" on the Internet. We believe that such interactions are an effective and efficient replacement for face-to-face gatherings, but they tend to lack the heart-to-heart perspective that is the essence of the genuine live meetings.

Internet connections such as webinars and summits serve a purpose, but they can never replace or come close to producing the result of truly live gatherings because online encounters lack a context of full presence and commitment. People often listen or watch while doing one or more other tasks and don't truly have their heart and soul in the connection. Accordingly, they (and everybody trying to participate) do not experience the full potential of *"swift awareness of the particular experiences of mutual interest."* And the result is that unique new ideas and thoughts are not spontaneously formulated.

Dozens if not hundreds of "Internet gurus" are now championing online events as critical to success, but these people are usually trying to make money and not make sense. They want us to believe that electronic connections can produce the same results as heart-to-heart and face-to-face interactions, but this is not true. Future generations may master the art of bringing their full selves to online gatherings. However, at this moment in time nothing can replace the face-to-face, heart-to-heart, mind-to-mind encounters of live meetings that do *"become pivotal in our lives."* We need to step away from the computers to come out and play. We need to hug each other so that we can feel the interconnection in our precious human bodies. And we need to experience the love and compassion that we share when we look into another person's eyes. This is the human experience, and it is only available when we come together with full presence in live meetings.

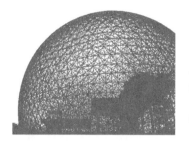

## CHAPTER 4

# DESIGN

## TRIMTAB ON
## THE PATH

**4.1** "The greatest of all faculties is the ability of the imagination to formulate conceptually. Artists have kept the integrity of childhood alive until humanity reaches the bridge between the arts and sciences."

EARLY IN HIS LIFE, BUCKMINSTER FULLER LEARNED THAT an individual can create anything imaginable and that the human capacity to imagine is one of our most potent capacities. People labeled as eccentric dreamers often become recognized as geniuses when their ideas and vision become popular, and that was true for Bucky as well as for people such as Einstein, Martin Luther King, Mother Teresa, Picasso, Frida Kahlo, John Lennon, and Emily Dickinson.

These visionaries initially were viewed as too far outside the mainstream to be considered acceptable and viable, but they persisted in following their imagination and dreams. That is the role of the artist regardless of the field in which they focus their attention and creation. It could be science or sculpture. It could be human rights or personal development. The possibilities are infinite, but the consistent thread flowing through the ideas and work of all these people is their integrity and willingness to sacrifice everything to follow their vision and to support the welfare of others.

These individuals may contribute works of music, graphic art, scientific discoveries, social paradigm shifts, or any of a multitude of possibilities. Regardless of the manifestation, it is always the result of passion combined with integrity. True artists in all fields of endeavor possess great passion, and they uphold their personal integrity as well as the integrity of their vision through thick and thin.

That is not to say that they are rigid. Bucky often reminded us, *"don't look to me for consistency, I'm always learning."* And his actions proved this to be true as he continued to have more "learning experiences" (what most people call mistakes) than his contemporaries. Still, he persisted in learning from his mistakes and continuing to do what worked. Following this simple path, he was eventually judged to be a success by society in general.

Still, his artistic imagination never ceased to shine. Even during the final year of his long life, he was still designing new structures and working with young people to build models of his ideas. And he always insisted that anything he created be beautiful. As he often reminded his audiences,

*"When I am working on a problem, I never think about beauty but when I have finished, if the solution is not beautiful, I know it is wrong."*

For Bucky and other "artists" in all fields, beauty is key to success, and he knew that it was the responsibility of all people who consider themselves to be artists to maintain the *"integrity of childhood,"* which he interpreted as *"the imagination to formulate conceptually."* Many of us might translate that phrase into the simple term "daydreaming."

Many ideas that were later judged to be a stroke of genius have arisen when a person focuses on an intention or a problem that needs to be solved and then allows her mind to "wander." During such times of imagining, the true artist will tune into solutions and new ideas and allow them to arise in a seemingly spontaneous manner. Taking the process one step deeper, she then uses the idea to morph something into existence—be it a new form of transportation, a book, a painting or a way to feed the starving people of the world. The form of the manifestation is not as important as the intention from which it arose and the context into which it will be manifest.

And it is these "artists" who maintain humankind's integrity of creation. They bridge the gap between what we label art and what we label science. Bucky labeled himself a comprehensivist, which made him an artist in many diverse fields. He consciously focused on bridging the gap between art and science, and he viewed them as different sides of the same coin. He also recognized that both scientific and artistic are creative endeavors in which people seek to uncover new realities and share them with others. They both require great creativity, a hallmark of Bucky's thoughts and actions.

Today, as we stand on the brink of a new era that Bucky said would either be "utopia or oblivion," each of us must employ our creative imagination to envision and establish a new paradigm of reality—Bucky's *world that works for everyone.* Just as Bucky reminded us that we are all *architects of the future,* he said we are also all artists and scientists of the moment, creating the survival and success of all sentient beings on Spaceship Earth with our every thought and action.

## 4.2 "Environment is stronger than will."

*Bobbi DePorter*

GUEST COMMENTATOR

These words of Buckminster Fuller have lived with me and grown in me for many years. In clarifying this statement he added ...

> *I would never try to reform man—that's much too difficult. What I would do was to try to modify the environment in such a way as to get man moving in preferred directions. ... I must commit myself to reforming the environment and not man, being absolutely confident that if you give man the right environment, he will behave favorably.*

How true and powerful Bucky's words are—we don't change man, we change his environment and man changes himself. That's exactly what Buckminster Fuller did in all his words and teachings throughout his life.

In the summer of 1981, I was part of a group that organized "A Week with Buckminster Fuller" at Kirkwood Meadows in the Lake Tahoe area. The space chosen was inspiring ... a lodge set in the mountains, a room with towering windows to take it all in. About a hundred people attended, each thrilled to spend time with Bucky and anxious to learn all he had to give us. Bucky had said he would share with us all he wanted the world to know from his work and life—a significant prospect indeed.

The group was positive, dedicated, and committed to learning. There was a sense of anticipation, joy, and sharing with each other, leading to an atmosphere of high trust and safety. The "environment" was set, orchestrated for inspired, engaged discovery. And Bucky captivated us throughout the week, every statement prompting a "what did he just say?"... we listened intently, we learned, we marveled, we contemplated.

His closing at the end of the week was even more meaningful and inspiring. Bucky was sitting on the stage surrounded by his artifacts and models ... slowly he picked one up, took it apart, and put it in his suitcase. Then a pause ... and another thought he wanted us to remember. Another pause ... and he packed another model ... then another comment. One after another he packed a model, paused, and shared a thought ... and we all took in every word. Finally Bucky had packed up all his models. He stood, picked up his suitcase, and walked off the stage and out the door.

We all sat in silence contemplating his words, reflecting on Buckminster Fuller. There had been no applause (Bucky didn't want any). If we had applauded, it would have changed the environment, and we would have given our power over to Bucky. In the silence we had to be responsible for all the learning ... it was ours now. Bucky created the environment and gave it to us—it was now up to us to learn from it and make change in the world. This was one of the most profound moments in my life.

That experience at Lake Tahoe was one of many I had with Bucky during the late seventies and early eighties. I have often reflected on the powerful belief he created in me about environment. There is no doubt in my mind that *"Environment is stronger than will."* And today I pass on that powerful belief to others in education. In our SuperCamp youth programs and Quantum Learning school programs we have proved over and over again—with teachers and with students—that environment is the key to creating a positive and effective atmosphere of learning.

Imagine a classroom with little order of any kind, from the seating, to the material on the walls, to the presentation of lessons. Imagine the kids in that classroom … the lack of attention and "order" there, too. Though the teacher tries to "reform" the students and make them learn, this environment is not conducive to learning.

Now imagine another classroom in which the environment has been considered. The classroom has been transformed into a "learning community" where every detail has been carefully orchestrated to support optimum learning … from the way the desks are arranged, to the use of music to cue desired responses, to defined classroom "policies," to the design of the lessons.

The atmosphere is one where students feel safe and supported and have a strong sense of belonging. The development of character-building life skills promotes respect and rapport— between the teacher and the students and also among the students. The tone of the class is comfortable and motivating. Every effort is acknowledged; all learning and achievements are celebrated. The kids make discoveries, learn and grow. The environment is conducive to learning—and they *want* to learn!

We don't change the students—we change the environment, and the environment changes the students.

BOBBI DE PORTER, president of Quantum Learning Network, was an early pioneer in the field of accelerated learning and its applications for effective learning and teaching environments. In the late seventies, she co-founded the Burklyn Business School based on the generalized principles of Buckminster Fuller. Bucky spoke at every school session until his passing in 1983. Bobbi applied the same learning methods to school-age children when she co-founded SuperCamp, a learning and life skills program with over fifty-eight thousand graduates in the U.S, Europe, Asia, and South America. The success of SuperCamp led to Quantum Learning school and community programs for administrators, teachers, students and parents, which has impacted over eight million students. Bobbi is the author of more than a dozen books on teaching and learning. She can be reached at: 800.285.3276 and www.QLN.com.

△ BUCKMINSTER FULLER WAS MASTERFUL AT TRANSFORMING environments. When he began his "56 Year Experiment" to determine what one individual could accomplish on behalf of all humankind, he had already come to realize that no individual can change another person's behavior, perception, or attitude.

Try as we might (and we do try), we can't change anything about another person. We can, however, change their environment and that environmental change will change their behavior. In other words, *environment is stronger than will.*

Successful educators and leaders employ this simple fact all the time, and Bucky was a master at both educating and leading. He never attempted to change another person or their beliefs. Instead, he worked to transform environments by inventing artifacts that made life easier for people.

He realized that people, like all elements of Nature, will always take the path of least resistance. Because of this he changed people's surroundings to make that path of least resistance an environment that led to a solution he had determined was in the best interest of everyone.

Because it is the most prevalent manmade structure on the planet, the geodesic dome is the most conspicuous of his environment changing inventions. It was not, however, created primarily for use as homes or playground equipment. It, like all Fuller's inventions, was invented to discover and demonstrate Nature's generalized principles. One such principle is the fact that the sphere is a perfect way to build and a model for a *"world that works for everyone"* because every point on a sphere is equal. The geodesic dome is the closest we humans can come to building a sphere using straight struts.

On a more practical level, the dome was created for two extremely important environment-altering uses. It was invented to provide temporary, inexpensive housing and to protect large numbers of people and huge amounts of resources from destructive forces such as severe weather and extreme temperatures. These two environment-altering uses clearly demonstrate how environment is stronger than will.

People may have a very strong will to survive, but a severe hurricane or earthquake can trump that will and instantly turn them into "helpless, hopeless victims." However, if they are under the protective cover of a geodesic dome when destructive circumstances occur, their environment will remain stable and so will their lives.

If people are rendered homeless and all their material possessions destroyed by natural disasters, flying lightweight, easy-to-assemble cardboard geodesic domes to use as temporary homes will alter their environment. In this instance, the environment-altering invention provides stability as well as a basic foundation for recovery.

In both instances, people follow the path of least resistance, either staying inside the huge dome or accepting the temporary domes as homes. And this is just one instance in which Bucky demonstrated the practical nature of working to change environments rather than trying (always unsuccessfully) to change people.

Bucky effectively employed this seemingly simple phenomenon throughout his own life to create changes that continue to influence us today even though we are not aware of them. All his inventions and artifacts (books and other writings, recorded talks, archive, etc.) were designed to change people's environments, thereby resulting in a positive change of behavior.

For example, rather than go out and proselytize to end global starvation and hunger, he wrote and spoke about the truth that in 1976 we reached a point in time when there was enough food on Earth to feed everyone. Educating people to that simple fact changed their environment because they no longer believed that there was insufficient food on Earth.

Before learning that fact, people thought that war was valid because everyone had to fight for their "piece of the pie," and they believed that there was not enough "pie" to feed everyone. Once their environment had been altered through learning some simple facts, they would soon begin to realize how important it was that "we the people" shift the focus of our resources from weaponry to livingry and begin to support all life on Earth.

Following Bucky's model for effective and efficient action, each of us can in fact make a huge difference in the lives of our fellow crewmembers on board Spaceship Earth. We can begin by simply resisting the temptation to try changing people or their behavior and look to see where we can change their environment to create more positive behaviors.

## 4.3 "Revolution by design and invention is the only revolution tolerable to all men, all societies, and all political systems anywhere."

⚠ MOST PEOPLE REALIZE THAT SOME TYPE OF MASSIVE GLOBAL change is underway, but they don't understand that the shift cannot succeed using antiquated behaviors, systems, and structures. We humans simply cannot continue to follow the path of recent generations and expect to survive—much less thrive. We have to mirror Nature and quickly return to being a sustainable species and culture. That sustainability requires the activity Bucky labeled a "design science revolution."

Bucky realized that the military could no longer dominate the latest technology and that our best engineers and scientists must shift their attention from weaponry to livingry. This is the design science revolution he championed. Bucky's inventions were livingry devices that supported sustainable life on Earth. He even labeled himself a "comprehensive, anticipatory design scientist."

He devoted most of his life to advocating for a design revolution that encompassed all humans, societies, and institutions. He campaigned for a world that works for everyone, and he believed that the only way that this would happen was if unique ideas were expressed as sustainable designs.

He was not, however, talking about what most people regard as design. Bucky was working well beyond that former notion of designing a building or a car using the same basic template as had been used for decades. Instead, he was looking into the future to find solutions for problems that had not yet arisen and taking the design of whole systems into consideration. Today, whole systems design is a popular field, but back when Bucky began the concept was essentially unknown.

Whole systems design requires comprehensive thinking. The most important physical whole system for humans is our fragile Earth. We must design for that whole system within the more comprehensive metaphysical whole system that we cannot experience with our physical senses.

When we design and create something, we need to consider the whole system of Earth in both the present and future just as Bucky did. By working in this manner, all creators/designers/leaders can begin to shape a sustainable environment for future generations. This is the revolutionary design science that Bucky predicted would support the survival and success of all humankind, and it offers us numerous opportunities for success both globally and as individuals.

### 4.4 "We are powerfully imprisoned in these Dark Ages simply by the terms in which we have been conditioned to think."

*By Zoe Weil*

What are the terms in which we've been conditioned to think? One of the most pervasive and perilous of mental conditionings is the ubiquitous either/or. We are conditioned from a young age to think in dualities. Things are black or white. They are as different as night from day. We are expected to be members of one of two political parties: Republicans or Democrats.

We argue one side or another of what should be considered complex challenges to solve, as in "loggers versus Northern Spotted Owls" or "do we invade/go to war with _____ [pick a country] or not." Our schools all have debate teams in which bright young adults face off around arbitrary and contrived either/or scenarios, such as these actual debate questions: "Is the U.S. responsible for Mexico's drug wars?" or "Are teachers' unions to blame for failing schools?"

What this conditioning does is not only prevent us from seeing grays and appreciating the spectrums on which we live, but also, more significantly, thinking as what I call solutionaries, whose goal is not to argue which side is right but rather to come up with answers to complex challenges and create a restorative, healthy, humane world for all.

As a humane educator and the president of the Institute for Humane Education (www.HumaneEducation.org), I believe that we can move beyond these dark ages and actually solve our interconnected global challenges by addressing one primary system: schooling. If we were to adopt a new purpose

DESIGN: TRIMTAB ON THE PATH 123

for schooling—to educate a generation with the knowledge, tools, and motivation to be conscientious choice makers and engaged solutionaries for a better world—our graduates would enter their chosen professions understanding their role in ensuring that the systems within them (whether in business and finance, defense and protection, energy and transportation, engineering and architecture, farming and water procurement, production and construction, politics and law, fashion and beauty, etc.) are sustainable, just, and peaceful.

They would do this as a matter of course because this is what they would have learned in their basic education from Kindergarten through college and graduate school. They would have been conditioned to think of themselves as responsible for contributing to a healthy world.

There are many ways to go about putting this idea into practice. For example, instead of the unhelpful task of arguing and winning a "side" in a fabricated either/or debate, we could invite our creative and bright youth to explore ideas for ending drug wars or improving schools. We could still have teams compete (if we believe competition is a positive way to create new ideas) to come up with solutions, but the goal for the teams would be to delve deeply into a complex problem and come up with the best, most innovative, most cost-effective solutions to pervasive and seemingly intractable problems.

We could also bring relevant information for solving current challenges into all courses and curricula. We could, in fact, transform the curricula we teach, recognizing that the "basics" of verbal, mathematical, and scientific literacy are simply foundational tools for living healthy, productive, contributing lives.

Each year of high school might have an overarching subject such as "Food and Water," "Protection and Conflict Resolution," "Products and Structures," and "Energy and Transportation." Students would learn the "basics" in the course of addressing these core systems (without which none of us could survive), and those "basics" would be ever more relevant and meaningful because students would understand their applications to a real world that is still replete with challenges that need to be addressed to create peaceful, healthy societies.

Thoreau once said, "There are thousands hacking at the branches of evil to one that is striking at the roots." Education is the root, and if we were to adopt a greater purpose for schooling—to graduate a generation of solutionaries—this generation would, quickly and inexorably, solve the myriad, interconnected challenges we face in creating a healthy, just world for all.

ZOE WEIL is the president of the Institute for Humane Education (IHE) www .HumaneEducation.org, which offers the only M.Ed. and M.A. programs in comprehensive Humane Education linking human rights, environmental preservation, and animal protection. IHE also offers online programs and workshops for teachers, parents, and change agents. She has given a TEDx talk on education and is the author of Nautilus silver medal winner *Most Good, Least Harm: A Simple Principle for a Better World and Meaningful Life; Above All, Be Kind; The Power and Promise of Humane Education*, and Moonbeam gold medal winner *Claude and Medea* about seventh graders who become clandestine activists in New York City. Zoe blogs at www.zoeweil.com. Find her on Facebook and follow her on Twitter at ZoeWeil.

△ A KEY ELEMENT CONTRIBUTING TO BUCKMINSTER FULLER'S success was his extraordinary ability to think for himself. Although society attempted to indoctrinate him into thinking and acting like a "productive, good citizen," Bucky was adamant that he and all people need never accept the mandates of others—especially those who are in positions labeled "leaders."

At the beginning of most of his marathon thinking-out-loud lectures, he would caution his audience not to believe anything that he (or anyone else) said. He would then go on to advise that people consider the ideas that others posed against their own personal experience. After doing that, a person can either accept or reject the ideas and act accordingly.

Bucky also often shared the things that he had discovered regarding the words we use and how limiting they can be, and he would coin a term to describe something that he found to be important that had no name. For example, while scientists were constantly telling people that the universe was entropic with all matter and energy traveling out and away, there was nothing to describe the significant phenomenon that is the opposite of entropy.

Bucky taught that although most scientists believe that entropy is the way of the universe, it has to have a corollary, a something where energy is absorbed inward rather than traveling outward. This phenomenon he labeled "syntropy"— a word he created by combining synergy and entropy.

Some scientists now believe that this phenomenon is what they label a "black hole." When this occurs in Universe, all energy and matter is sucked in, so nothing can be observed using our physical senses.

The other place that Bucky observed this phenomenon of synergy happening is on Planet Earth. Energy and a small amount of matter in the form of space dust are constantly coming in to our planet. Then, that energy coalesces even more as plants turn it into physical matter. And with that creation, we have a word to describe a most important aspect of reality and a phenomenon upon which we depend for life.

This is only one example of the ways we humans keep ourselves in the "Dark Ages" with our words. Another is a word that you just read, and that Bucky did his best to eliminate from his vocabulary. That word is "believe." Bucky would tell his audiences that he did not believe anything. He would say that "either I know it or I don't," but he did his best not to use the word believe and to look at all aspects of his life to see what he knew and what he did not.

By making that distinction, we are forced to do what Bucky chided us to do and not believe anything we see, hear, read, or are told is true. Rather, we can compare these external aspects of our reality against our personal experience and determine what is true for ourselves. That was the Fuller way, and it served him well for decades.

**4.5** "The most important thing to teach your children is that the Sun does not rise and set. It is the Earth that revolves around the Sun. Then teach them the concepts of North, South, East, and West, and that they relate to where they happen to be on the planet's surface at that time. Everything else will follow."

*By Ann Medlock*

GUEST COMMENTATOR

*Eppur si muove*—"And still it moves"—said Galileo, centuries ago when the Inquisition forced him to retract his writings about the Earth's movement around the Sun. And yet we go right on talking as if the Earth were the center of the universe, feeding the ignorance of those who actually believe it *is*. Buckminster Fuller couldn't stand that.

I heard him do a variation on this theme at his last public appearance in New York. He told us that he'd asked from the podium at a conference of scientists, *"How many of you saw the sunrise this morning? How about the sunset yesterday?"* He'd looked at the many raised hands and in effect said, *"Shame on you."* They were scientists. They knew the Sun neither rises nor sets.

It was hard not to feel sorry for those entrapped hand-raisers. We've all "seen" sunrises and sunsets and we all know that's not what we're seeing and yet we all say those words, right? You *could* see it as an innocent recourse to the poetic, but what Fuller saw was the willful spreading of misinformation, the denial of what is real and proven and scientific. It bugged him.

Being a writer, even sometimes a poet, I'll probably never say, "Didn't the atmosphere look gorgeous last night when our position on the planet rotated away from the sun?" As a parent and as a creator of classroom materials on compassionate, courageous involvement in the world, I don't agree with Fuller that helping kids understand the movement of the planet they're standing on is *the* most important thing to teach them. But it's a biggie.

It's a dose of reality, a core lesson in proprioception, in nonsense-detecting, and in calling a spade a spade. Given the alarming numbers of people who live by multiple fictions and are indeed willfully denying proven realities, debunking sunrises and sunsets could be a great foundation for introducing kids to the world they actually live in. Living based on facts rather than fictions—what a concept.

As another wise human once said, we're entitled to our own opinions, but not our own facts. Fuller lived, breathed, and taught on the solid ground of fact, ever the scientist, backing up his spectacular imagination with proofs, with working, operational models that dazzled the world.

ANN MEDLOCK is the Founder and Creative Director of the Giraffe Heroes Project (www.giraffe.org) and author of *Arias, Riffs & Whispers,* and *The Mermaid's Tale.* For more information about her writing go to www.annmedlock.com.

⚠ THIS SIMPLE TRUTH WAS EXTREMELY IMPORTANT TO Buckminster Fuller because it was about view and perspective. We've all been indoctrinated to believe that the Sun rises and sets—a perspective that implies everything revolves around us and that each of us is more than a single aspect of Universe. Building on that perspective, many of us living in the developed nations have continued to behave as if we are the center of everything and entitled to anything we desire with little or no regard for the welfare of other sentient beings.

This view does not respect or honor our beloved Gaia—Mother Earth. It also tends to keeps us from being thoughtful, compassionate crewmembers onboard the planet Bucky named Spaceship Earth and from recognizing that we are stranded on a fragile, lifeboat planet.

When we preach this gross lie to our children, we continue to propagate a perspective that is destroying humankind and other species. We also continue to champion the belief that humans are superior beings who can do anything we want and continue on an unsustainable path. And we instruct future generations to keep on doing what we have been doing with the irrational hope that it will finally work.

We instill in our children and grandchildren the belief that they can take all that they want without regard to others at a time when many of us have realized that we are a single species who must work together and follow the dictates of Nature. We're all in this together, and we need to cooperate and come from love rather fear if we are to survive and thrive as a species and as individuals. This is the perspective of realizing who you are and where you are in physical space and in relationship to all Universe.

Bucky championed the fact that knowing where you are in relationship to the elemental surface of our planet is essential to our development and ability to contribute to others. If a person has little or no orientation on Earth, and in relation to other people and physical reality, he can't proceed forward (on any level) because he does not know where he is or what possibilities exist. In other words, you can't know where you are going until you know where you are.

This is true in both physical reality and metaphorically. We need to wake up and appreciate where we are and what is happening around us. Only then can we take sustainable, positive steps in a direction to benefit ourselves, our families, our neighbors, and all sentient beings. Only then can we begin to step up as truly global citizens. And only then can we experience the love, peace, and joy that is our natural birthright.

The primary reason that this perspective is not taught or experienced by most people goes back to the time when "the church" seized power and became the established authority. This was a period of transition from a natural spirituality that was in harmony with Gaia Mother Earth to an era when a small group of men exerted their authority as religious leaders and promoted their own welfare over that of all others.

These men would have us believe that the Earth is flat, and for centuries they challenged anyone who argued otherwise. And even though we know that the Earth is spherical (not round, which is still only two dimensional), most of us subconsciously feel that we live on a flat surface and, thus, we say that the Sun rises and sets. This subconscious feeling allowed that handful of religious leaders (and now primarily leaders of our new religion—commerce and business) to control the majority of our resources and maintain the status quo.

Their perspective of a flat Earth established the directions "up" and "down" with "up" being good and represented as heaven and "down" being bad and represented as hell. At a time when resources were scarce and life was hard, religious leaders taught people that their next life would be better if they were good citizens, followed the dictates of the church and got into heaven. "Up" and heaven can only exist if we believe that our reality is two-dimensional and flat.

Those same religious leaders also empowered themselves to decide who went up and who went down. Thus, people obeyed the mandates of those greedy men and did what they were told in order to support organized religion. Today, we continue this tradition, but the primary forces of greed are business and government leaders.

On a spherical Earth, we are all equal. There is no point that is more important than any other on a sphere. There is also no up or down, heaven or hell controlled by a small clique of elite people.

Bucky Fuller's view is multi-dimensional. On our spherical Earth there is no up or down. Everything exists in relation to everything else, and the correct words are "out" and "in." With that perspective, everyone has an equal share of power and control, and we begin to establish Bucky's vision of *"a world that works for everyone."*

To solve the issue of sunrise and sunset, Bucky coined new terms that reflect our experiential reality. These terms are sunsite and sunclipse. These words communicate what really happens at those times. In the morning we site the Sun as it appears on (not above) the horizon, and in the evening it clipses as it disappears on (not below) the horizon. Were we to start using those terms and teaching them to our children, we would begin to live in a more honest and inclusive reality where each of us is of equal value to the whole.

## 4.6 "Faith is much better than belief. Belief is when someone else does the thinking."

*By Satyen Raja*

GUEST COMMENTATOR

Belief is a young person's first software by their society and surroundings.

Belief is the beginning scaffolding that youth ascends (regardless of age) to gain some ground on the unknown.

Belief can be a placebo that actually works even without rationale.

Belief is the rubber band ball that allows little in (even if you think so).

Belief hits its ceiling when beliefs don't match up with what's in front of you.

Faith is the hope-spring rooted in the invisible.

Faith is the rallying call of mad leaders.

Faith can be the fallacy of being saved from above from your outs with integrity.

Faith is the relaxing of tension and striving to forces beyond our comprehension.

Faith also hits its ceiling, as it's prone to delirium at times ...

Then there's the next better step ... Knowing.

Knowing is the fullest maturation of Faith through the most exhaustive honest seeing through of your beliefs.

Knowing is earned from the fire of experience, earned by conscientious (and sometimes silly) trial and error.

Knowing comes and makes itself known when one settles down into what is Real for themselves.

Knowing doesn't need an outside thing to have Faith in or an inside set of ideas to believe in.

Knowing is irrefutable and isn't based on facts or perceptions that can be viewed from different angles.

If it can be viewed from different opinions, dig deeper to the place that transcends any differences.

This takes knowing that Knowing is worth Knowing and much richer and freeing than what anyone else can give you.

Faith and Belief can get one to kill with patriotic zeal.
Seeking real Knowing shines light through all of it!
Knowing is your best defense and the last stand.
Knowing too hits its ceiling, even with the most experienced of us.

And the most experienced of us know there is that we can never know … The Unknowable.

SATYEN RAJA combines the power of the Warrior and the wisdom of the Sage to inspire audiences worldwide. Satyen is a unique blend of power and of heart. He has taught tens of thousands of students around the world the art of 'true power.' Satyen's dramatic style is not for the timid. Through his teachings, your weaknesses become strengths.... Quickly! He is the founder of The Get A Life Company and can be contacted through www.TheGetALifeCompany.com.

BUCKMINSTER FULLER WAS CONSTANTLY ADVOCATING THAT everyone think for herself or himself. He often told his audiences, *"don't believe a word I say or what anyone else says or writes."* He would go on to urge people to be moved by child-like curiosity and fully consider everything that comes into their environment. Then, one needs to compare that incident or thing against one's personal experience to see if it feels valid.

If it does, a person can accept it, but if it goes against a person's experience, core values and integrity, an individual needs to reject it—even if the behavior or object is part of accepted society. That is what thinking for yourself is about.

Bucky defined belief as allowing others to do the thinking for you and blindly accepting their mandates as the truth. He said that faith in a "Greater Intelligence" (Source / Great Spirit / God / Allah / Higher Power, etc.) was a healthier way to operate, but he also advocated thinking for yourself above even such faith.

His personal discipline in this area involved language. Thus, he did his best not to use the terms "believe" or "belief." He would tell his audiences, *"either you know something or you don't."*

This may seem like a matter of semantics, but if you stop to consider what you are saying and advocating when you say that you believe in something or believe something to be true, you may notice a shift in your thinking. I began doing this many years ago, and I have found it to be a very useful technique for determining what is true for me and what is the result of outer indoctrination.

For me, the often-taught prioritizing process when a person divides everything into three categories can be helpful here. "A" are the important things. "B" are the things that may be important. And "C" are the unimportant things that can simply be passed over. Beliefs often fall into the "B" category.

The second step in this process is to go through the "B" pile and put these things into one of the other categories. Thus, you have things to complete or keep and things to forget or toss out.

Using a similar exercise with what you believe will result in things that you know and things that you don't know with nothing in between.

One interesting thing that Bucky found he knew as he got older was how little he knew. In fact, he wrote an epic poem entitled *How Little I Know.*

This is the perspective of a true visionary and genius. They recognize that Universe is infinite and beyond our knowing. Even though they are often glorified as brilliant and insightful, they are humble enough to know how little they actually know, and they eschew such accolades.

They realize that there are no beliefs but rather data. And even that data and seeming facts are subject to interpretation and change depending on the perspective of the observer. Thus, it is better to have faith than to believe in something. It is, however, even better to know or know that you do not know and to constantly evaluate what society and others would have you believe against your personal experience.

### 4.7 "How often I found where I should be going only by setting out for somewhere else."

*By Roshi Joan Halifax*

GUEST COMMENTATOR

I met Bucky numerous times over the final years of his life. His visionary capacities and "beginner's mind" made every territory that he entered new—and new for those who were fortunate enough to be with him. He had a curious way of tilting his head when observing or listening, as though he wanted to see things from a totally different perspective. He was a failure gone right, a loser who found truth, a nonconformist who brought new forms into being.

He was also dedicated to ending suffering in this world, as someone who had suffered loss and humiliation through much of his early life. In the late 1920s, he dedicated himself to *"the search for the principles governing the universe and helping advance the evolution of humanity in accordance with them ... finding ways of doing more with less to the end that all people every where can have more and more."* Now, if these aren't words that relate to our era, I don't know what other words would suffice!

I have always cherished my time with Bucky. He was an inspiration for me. He was out of the box and into this world. He was a systems person, saw the magic of emergence, saw how all beings and things were inextricably connected. He thought globally, was a dedicated environmentalist, and was completely in touch with the local effects of our consumption. He once said: *"There is no energy crisis, only a crisis of ignorance."* As a Buddhist, I have long known he was right: our ignorance (supported by our greed and aversion) is bringing immense suffering into our world. Bucky was dedicated to finding a way through this in action and intuition.

In the end, Bucky knew he was verb, not a noun. His whatness was not his greatness. His greatness was found in his ability to live with ephemerality, to live with the truth of change.

ROSHI JOAN HALIFAX is a Buddhist teacher, author, social activist, and anthropologist. She is Founding Abbot of Upaya Zen Center, a Distinguished Visiting Scholar at the Library of Congress, and has written books on Buddhism, religion, and end-of-life care. She is a pioneer in the end-of-life care field and engaged Buddhism. Learn more about her work at www.upaya.org.

*By LD Thompson*

GUEST COMMENTATOR

Buckminster Fuller was a man of humble genius. Though celebrated as an innovator and visionary, he was first to acknowledge that there are those times when, setting off in a direction, something entirely else can happen and course correction is necessary.

I have seen this acted out in the lives of many talented people I've worked with through the years. One set off to be a musician and found himself becoming a computer programmer. Another set off to be a teacher and wound up designing websites. Several planned to be entrepreneurs and found themselves behind a desk working nine to five.

Do these changes, these course corrections, represent failure? Or do they represent something sublimely providential—communication from our deepest intelligence, guiding us surely and wisely toward the life that we need most to live, adhering to a mandate for growth that is our prime directive?

If it were possible for each of us to eradicate the concept of failure from our vocabulary, it would be so much easier for us to listen to that deeper intelligence, our Soul, and be guided by it. But as human beings we tend to get caught up in our conditioning and live out of the moment sometimes hanging onto old dreams long after their pertinence in our lives.

Buckminster Fuller loved nothing more than being out on his sailboat, the *Intuition*, riding the wild waves of this fascinating life. He once offered a valuable metaphor:

> *"On the edge of a large ship's rudder is a miniature rudder called a trimtab. Moving that trimtab builds a low pressure which turns the rudder that steers the giant ship with almost no effort. In society, one individual can be a trimtab, making a major difference and changing the course of the gigantic ship of state."*

The "American Dream" is experiencing a sea change. In the midst of an historic economic downturn, the majority of Americans are having to adjust their financial expectations and their capability of achieving their goals. But something greater, more refined, can come out of these adjustments. Those whose lives are dedicated to being awake and aware are finding that their lives are even more meaningful with greater integrity and humility being demanded of them in this moment.

I know several people who have lost a significant portion of their financial net worth, but they are happier now and more alive than when they were swimming in equity.

As a result of not resisting our Soul's curriculum for our life, we are at the point of truly finding out where we should be going ... after having set out for somewhere quite different.

Each of us, during this sea change, has the opportunity and the responsibility to be a trimtab. Every moment, every act, every thought that contributes to consciousness and creativity is potentially that adjustment necessary to turn this ship of state toward a more sustainable direction.

Every choice that values community, compassion, nonjudgment; every thought that acknowledges that we all live on the same planet and breathe the same air and drink the same water; every innovation in an individual life that offers respect for all life is an act that acknowledges that we, as Souls, are born of the same stuff—the same material, the same intelligence and, ultimately, serve the function of the trimtab.

This course correction has the potential to deliver us to a very new land, where the Soul's values—integrity, honesty, generosity—are the values by which we live and express our creativity in our world.

LD THOMPSON is an author, inspirational speaker, filmmaker and social activist. Having spent over twenty-five years exploring the mind/body/spirit connection, LD works with groups and individuals around the world, helping them find their passion and power through spiritual practice. His new book, *THE MESSAGE: A Guide To Being Human,* is about how we can understand our "Soul's Curriculum." Learn more at www.divineartsmedia.com

△ IN THIS QUOTE BUCKMINSTER FULLER ELUCIDATES THE wisdom of elders that so many of us have learned over the years. We all set out in life with goals and visions only to find that they take us somewhere completely different. Universe and our past actions (sometimes labeled karma) send us in particular directions that are unseen at the onset of the journey.

This book and my relationship with Bucky represents yet another such journey. *A Fuller View* started nearly ten years ago as my attempt to share more of Bucky's insights and wisdom online. I would publish a quote followed by my explanation. Eventually, that became a regular feature on my Facebook page, and then morphed into longer writings after I reconnected with the book's publisher Michael Wiese on Facebook.

At that point, this was simply an interesting book designed to share Bucky's ideas and my perspective. Then came the idea of adding "guest commentators," and that led to the unfolding of a whole new set of possibilities and challenges in order to be of greatest service to the most people.

Bucky, too, had many such unfolding, invigorating experiences. As he often reminded his audiences,

*"I didn't set out to design a house that hung from a pole, or to manufacture a new type of automobile, invent a new system of map projection, develop geodesic domes, or Energetic-Synergetic geometry. I started with the Universe ... I could have ended up with a pair of flying slippers."*

After years of trying to steer the ship of his life himself, he finally realized a much bigger "something" that he labeled "Greater Intelligence" governed all Universe, and that it would be much easier to begin with as broad a perspective as possible and allow the flow to take him where he would be of greatest use.

That "somewhere else" is usually far more interesting than where we set out for—because it's almost always an unknown.

Certainly, we can try to remain within our comfort zone, but the Greater Intelligence (within us and around us) will take us to places and situations that may be uncomfortable and support of our growth and development. This is where our true self wants to be, even when our lesser self vehemently resists.

Bucky made a practice of being aware of that unknown and enjoying the journey. He often did not know where he would be led, but he made a point of saying "yes" to invitations whenever possible.

This is a challenge we all face, and it is particularly difficult at this moment in human evolution when we are morphing from a warring competitive culture to a society predicated on peace, love, and joy for all. Saying "yes" to opportunities and challenges that seem to lead in directions and to places that seem to be out of sync with society is not easy for most of us. Still, "yes" is the required answer if we are to survive and thrive as a species.

And, in the end, saying "yes" to the unknown will bring us to a higher standard of living and loving than has ever been known to humans. We just don't know exactly what that new reality will look or feel like yet, but many of us have an intuitive sense that it exists and we're excited about the unfolding of this possibility.

# Start With Universe—End Up With Contribution

## Chapter Conclusion

**4.8** "I didn't set out to design a house that hung from a pole, or to manufacture a new type of automobile, invent a new system of map projection, develop geodesic domes, or Energetic-Synergetic geometry. I started with the Universe—as an organization of energy systems of which all our experiences and possible experiences are only local instances. I could have ended up with a pair of flying slippers."

WE ARE ALL DESIGNERS OF THE FUTURE—BOTH OUR individual future and our collective cultural future. Even though most of us don't often think about the broader aspects of our lives or believe that we're subject to the dictates of society and leaders, we are the masters of our own destiny as well as the destiny of future generations. Each of us does in fact create our own reality, and we need to become more conscious creators.

The issue is not what we want or what we do, but rather what is our intention and the scale of our perspective at the onset of creation. Ordinarily, we simply decide that we want to have or do something, and we begin the process. We don't really consider the big picture and how we can be of the greatest service to the most people. We don't think about how our desires will affect other people or aspects of Universe. And we certainly don't take the thoughts and feelings of others into account. We simply know what we think we want, and we begin to create it.

This way of operating was not always the predominant method, and a more communal broad perspective continues to thrive in a few small isolated pockets of civilization. People in these places live in community, and they consider the environment an essential aspect of their life support system. Thus, they would never chop down a forest of trees simply because they needed the wood for heat or the land for farming. They recognize that the trees and all living organisms are part of a greater whole, and to survive and thrive we must honor all life.

Knowledge of the "real" world that Nature has created is extremely important to them just as it was to Buckminster Fuller. Following the 1927 epiphany that began his "56 Year Experiment," Bucky shifted his design/creation focus from a self-centered model to a Universe-centered model. In other words, he began working with as large a vision as possible. For example, he did not initiate his Dymaxion Vehicle project by deciding to design an advanced car that he could mass market. He started with the whole of Universe and from there looked to see what problems would present the largest challenge to people living fifty to one hundred years in the future.

A major issue he envisioned as a challenge was human transportation, and from a broad perspective Bucky realized that we could not depend on fossil fuel forever, as was the common belief when he initiated the Dymaxion Vehicle project in the 1930s. Thus began the emergence of what he would eventually label an "Omni-Medium, Twin Jet Stilt Transport" that could be used as transportation on land, on water, or in the air. Although Bucky's young daughter Allegra labeled the model for this vehicle a "Zoommobile," it was not some fantastic science fiction apparition. The Dymaxion Vehicle was in fact the first phase of a transport that Bucky envisioned being viable once the technology that he predicted would happen did emerge.

"Real world" issues that most people had to face during the Great Depression were not issues for Bucky when he set out on this quest. The Dymaxion Vehicle was designed to solve many of humankind's future transportation problems, and if it were mass-produced today, it would provide viable solutions. This is what happens when a person begins with Universe and designs from a broad perspective in order to support all sentient beings, and it is the antithesis of what most people and corporations have been doing during the past century.

Look at transportation today to witness the opposite of starting with Universe design. The automobile and the internal combustion engine emerged from a very narrow vision and a desire to make huge profits for a few individuals and corporations. The "winners" created an impractical vehicle that consumes enormous amounts of resources and uses an engine that has thousands of moving parts and is designed to need lots of repairs and eventually wear out. There is no system to recycle any of the elements of that vehicle or its waste products, and thus the "modern" automobile and internal combustion engine is destroying our environment and may eventually lead to the extinction of our species.

Even at the time of the internal combustion engine's inception, there were other alternatives offered, but they were proposed by men who, like Bucky, were more interested in solving problems than making huge profits for themselves. In other words, their primary motivation was to make sense rather than trying to make money.

You and I now have an opportunity to shift this unnatural way of operating so that we begin with an intention to make sense rather than an intention to make money. This is the aspiration we must have at the onset of every new design project (and the creation of almost anything is a design project) if "we the people" are to take control of our planet, our lives, and the welfare of future generations. We must always "*start with Universe*," which includes the welfare of all sentient beings, if we are to survive and thrive on our tiny, fragile Spaceship Earth. The result might be "*flying slippers*," but the result could also be Buckminster Fuller's vision of "*a world that works for everyone.*"

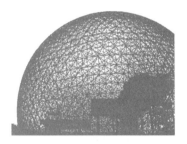

## CHAPTER 5

# CONSCIOUSNESS

## WAKING UP TO BEING AWAKE

# 5.1 "There is nothing in a caterpillar that tells you it's going to be a butterfly."

△ IN THE SAME WAY A BUTTERFLY EMERGES FROM A COCOON, we are a global society emerging to be something beyond what most of us can imagine. Our cocoon of collective ignorance and accumulated planetary resources has served us well through our development, but it is now time for us to emerge as genuine Earth stewards creating a sustainable world that serves all sentient beings.

Buckminster Fuller was one of the first people to recognize this forthcoming paradigm shift, and he did all that he could to live his life as an experiment and demonstration of the new emerging human while sharing all that he learned along the way. He was what we now label an early adapter or perhaps a visionary implementer who recognized the unity of all and lived his life in support of all sentient beings.

We can be truly grateful for Bucky and people like him who were—and still are—willing to take risks that are beyond what most people are even willing to imagine. He showed us what is possible when a person realizes that he does make a difference and then takes action on behalf of as large a group as possible.

Bucky knew that he would probably not eliminate global starvation, illiteracy, homelessness, and disease in his lifetime, but he continued to campaign for these and many other causes until he breathed his last breath. He also recognized that we really don't know what we are emerging into, but he did his best to predict what problems would be most challenging to future humans. Then, he created solutions to those issues for us—the future generations, many of whom he would never meet.

This is the model of the butterfly evolution that the human species is becoming. This is the next chapter of our blossoming, and there is no evidence in the behavior of past generations that gives any clue what will emerge, much less what will create success. This is, however, the only way we will survive and thrive as a species, and we are fortunate to have had a few people like Bucky to show us the way we can consciously transform our lives and our global society.

Prior to Bucky, many other people preached a similar message of abundance, love, peace, joy, and cooperation, but they were not popular and much less effective in changing things. These men and women were harbingers of the future, but they were usually overlooked or prosecuted for stirring up trouble. There were more such outlaw troublemakers in the past, but most of them were not in the public spotlight.

Bucky would have suffered that same fate had he not made a conscious decision to create his "56 Year Experiment" to discover and document what a single individual could accomplish in one lifetime. If he had not made a conscious effort to document his achievements (not because he had a huge ego, but to prove to us all that it could be done), this book would not exist and even fewer people would benefit from his modeling of how a fully responsible global citizen behaves.

Today, however, we know much more about ourselves and our environment, but we have no idea what our next stage of human evolution will look like. There is nothing in our past that gives us the slightest glimpse of who we are becoming. We are an emerging species, and Bucky recognized both the transformation and the unknowable nature of the outcome.

He did realize that our shift was primarily one of metaphysical consciousness that would then be become manifest in our physical environment. Because of this, he worked mainly with physical artifacts (inventions) to demonstrate the metaphysical principles that he realized would be key to our successful emergence into the light of a new era. From his focus, we can all learn a great deal, and for that focus we can all be extremely grateful.

**5.2** "Very very slow changes humans identify as inanimate. Slow change of pattern they call animate and natural. Fast changes they call explosive, and faster events than that humans cannot see directly."

⚠ CHANGE IS INEVITABLE AND ONGOING. NOTHING IS THE same as it was a moment ago. Even the page you are reading is changing and morphing, but that change is so slow that we label the page of a book as inanimate. Leave that book outside in the rain or too close to a fireplace, and the change becomes very obvious as the book disintegrates from exposure to water or fire. We continue to label the paper as inanimate be it in the form of a soggy ball or ashes, but it no longer appears as the page of a book.

If something is changing somewhat faster but still slow enough for us to see the transformation, we label that type of thing—be it a human or a flower—as animate. We say that it is living even though the physical matter that composes it is constantly changing, just like the physical substance of the book page.

All matter and energy is continually flowing; yet some things are considered living while others are not. All things are composed of the same atoms and molecules, but we make a clear distinction between living and not living, between animate and inanimate. Although this is a manmade classification, it does help us to define our reality.

Buckminster Fuller points out that we also classify faster changes such as an earthquake, a car crash, or a birth as explosive. We can see and experience these changes, but they are so dramatic that we label them as volatile, unstable, and unpredictable. They are, however, simply changes happening in a shorter amount of time. With a broad enough perspective, they are also predictable. Most humans do not, however, possess such an expansive perspective.

Those first three categories of change are filled with phenomena that we believe we have some control over. The last category, however, is composed of the changes that really matter, and we humans can't even see or experience them directly. These are in the realm of the phenomena Bucky referred to when he stated,

*"99.9% of all that is now transpiring in human activity and interaction with Nature is taking place within the realms of reality which are utterly invisible, inaudible, unsmellable, and untouchable by human senses. The invisible reality can only be comprehended by metaphysical mind, guided by bearings toward something sensed as truth."*

These are the *"realms of reality"* in which the majority of important action and activity occurs, and we don't even notice it. We humans believe ourselves to be so superior, but we can't comprehend most of Universe. We believe that we know how to manage our lives and our planet, but we're living in a preschool consciousness where we don't recognize some of the most obvious changes even though they're right in front of us.

How many times have you commented about a new picture in a friend's home, the color of your coworker's new coat, or your colleague's latest hair style only to learn that the change you just noticed took place weeks or months ago? And how many times have people complimented or criticized you about something they believed to be new when it was actually weeks old?

We're all guilty of being on both sides of this collective unconsciousness, but we don't realize just how rampant it is. We also don't realize that we really don't notice any of the important changes and that we're really not in control of anything.

**5.3** "Our children and our grandchildren are our elders in universe time. They are born into a more complex, more evolved universe than we can experience or than we can know. It is our privilege to see that new world through their eyes."

*By John Robbins*

GUEST COMMENTATOR

What that remarkable statement has continually given me is a reminder that the innocence of young people is not a fault to be corrected, nor an ignorance to be educated, but a strength to be celebrated. Sometimes it takes courage to be innocent. Sometimes the truly daring thing to do is to be naïve. Bucky's words remind me that my cynicism can be a trap, that my suspiciousness and defensiveness and know-it-all-ness can blind me, can keep me from awe, wonder, and amazement.

Of course it is our task to guide our children so that they may live with respect for themselves and respect for others. But Bucky reminds us that our function is never to seek to impose our limited and limiting beliefs upon them, but rather to help them to develop so they may give the gifts they are here to give, and to receive the gifts that others have for them, as well.

My son, Ocean Robbins, has been a superb teacher for me in this respect. He builds on all that I've accomplished. He and I are stunningly compatible in our values and visions. But he goes places I could never go. Like all our children, he is not here to fulfill my expectations, nor to follow any agendas I might have for him. No, he has his own dreams and aspirations. He has his own path to follow, his own sorrows to learn from, his own joys to be nurtured by, and his own love to share.

When I become too cynical, I forget that our children and our grandchildren are conscious in ways that are beyond me. They carry the seeds of the future. They come into the world endowed with new possibilities, new understandings, and new energies.

If we can align ourselves fully with them and learn from them, they may be able to accomplish things we cannot. They may be able to correct the errors of past generations, including our own. If we can truly live by the wisdom Bucky is offering us, our children will not only be free to be themselves. They will bring a new breath of life into the world.

JOHN ROBBINS' work and information can be found on his website www.johnrobbins.info.

*By Ocean Robbins*

GUEST COMMENTATOR

I see this quote from the perspective of both a youth and an elder. As a youth, I grew up with parents who believed that I had a purpose and a perspective that they could support and learn from, and that their job was to help me to give my gifts— not to force me to conform to some expectation they might have held for me. My dad told me time and again that he was proud of me for all my accomplishments but that he would love me just as much if I were autistic.

So I felt a deep knowing that I could be me, whatever that turned out to be, with whatever foibles and brilliance might emerge, and that my parents were utterly devoted to supporting and appreciating the manifestation of my unique insights and true path. I can say from firsthand experience how precious it is when elders truly see and stand for the unique gifts and journeys of each new generation.

Today I am also a dad. My identical twin boys, as it turns out, actually are autistic. They have their own unique brilliance and their own very significant struggles. And I have the opportunity, now, to love them just as they are, with all my heart.

My children's wisdom sometimes doesn't show up in forms that are readily accessible to me. But when I let go of my hopes and fears, my ambitions and my goals for them, and when I settle into the deep truth of what is so, I am filled with an appreciation that is ineffable and eternal—and that changes my life.

So as much as I might want to teach my children all the things that matter to me and to transmit my values and insights, I am learning how much I gain from being present with what matters to them. I am finding a sweet depth of connection that we can share when I let go of trying to drag my children into my world and instead accept their sweet invitation to be a companion in theirs. In this process, I am learning about real love, about release of expectations, and about a quality of presence that is changing my life.

As a mentor to thousands of young leaders worldwide, this quote inspires me to reflect that in a constantly changing world, many of us have opinions, beliefs, and modes of analysis that function rather like a computer's operating system. These help us to interpret the overwhelming array of information that comes to us, and to filter, digest, and determine appropriate response to stimuli.

Many of us, however, are interacting with the world using operating systems that were built when we were kids or even that were passed down from previous generations. We tend to settle into familiar habits and over the years can become increasingly inflexible. We need constant updates to our operating systems in order to stay present with a constantly changing world, as well as with our constantly evolving selves.

Children and young people are often not only more adapted to the world into which they have been born but also more open to updates as they take in new information or learn to perceive in new ways. This porosity and adaptability has much to teach all of us. I believe that growing up doesn't inherently cause us to lose flexibility, but that if we don't keep stretching and changing along the way, it gets harder and harder as we move through life. In this, young people can always be our teachers.

OCEAN ROBBINS is author of *Choices For Our Future* and *The Power of Partnership* and founder of Youth for Environmental Sanity (YES!). He has facilitated gatherings for thousands of leaders from more than sixty-five nations, is a recipient of many awards including the national Jefferson Award for Outstanding Public Service, and speaks widely at conferences on university campuses. To learn more, visit www.oceanrobbins.com.

*By Lynne Twist*

GUEST COMMENTATOR

Buckminster Fuller, or Bucky as he was affectionately called, was one of our children's favorite dinner guests. One evening at dinner our daughter, Summer, who was around eight years old at the time, made a profound comment. I'll never forget Bucky looking at my husband Bill and me saying, *"Never forget that our children are our elders in Universe time."*

From that day forward I saw everything differently: my children's deep wisdom, my role as their mother, and my personal responsibility for future generations. It was particularly poignant when my first granddaughter, Ayah, arrived. Born in 1999 at the end of the twentieth century, I realized she could potentially inhabit three centuries in her lifetime, given female life expectancy trends. I can't imagine what she will witness and experience in her lifetime.

Ayah and her brothers come from a rich ethnic and cultural heritage that is so different than the family in which I was raised. My grandchildren are being nurtured as global citizens by conscious, loving parents who already knew what I learned in my 30s: that our children and grandchildren are born already knowing and learning at warp speed the mastery of complex technology and sophisticated communication tools. This *is* the order of the Universe and it is incumbent upon us to learn from younger generations as well as to share the wisdom gleaned from years of experience.

Besides being a grandmother, one of my greatest joys is to inspire, empower, and create shared learning opportunities with the millennial generation. The current generation of twentysomethings blows me away with their commitment to making a difference with their lives, accompanied by their intelligence, a spirit of "Yes, we can," and their intuitive knowing that we only have one planet and we are one human family.

Lynne Twist is Co-Founder and Board Member of The Pachamama Alliance. She is author of the award-winning book, *The Soul of Money*, and President and Founder of the Soul of Money Institute. Lynne is also a global activist, fundraiser, speaker, and consultant, who has devoted her life to service in support of eradicating hunger and poverty, empowerment of women and girls, global sustainability and security, human rights, economic integrity, and spiritual authenticity. She has raised hundreds of millions of dollars and trained over thirty thousand fundraisers worldwide to be more effective in their work with social profit organizations. For more information about her work, please visit www.soulofmoney.org and www.pachamama.org.

## By Marilyn Schlitz

GUEST COMMENTATOR

Bucky's life was dedicated to the service of humanity. As an iconoclast, he challenged assumptions that many take for granted. One area of which he was highly critical was mainstream education, where the emphasis often is on rote learning rather than creative problem solving. He once noted: "*All children are born geniuses. Nine hundred ninety-nine out of every 1,000 are swiftly and inadvertently degeniused by adults.*"

As a visionary and social reformer, he argued that there is a growing need for the development of skills and competencies that will allow students to successfully meet the challenges of our changing world. Bucky felt strongly that in addition to enhancing their academic performance, students need to master new skills that will prepare them for their place in the evolving universe. Being able to adapt to changing circumstances is essential for students who are growing up in today's complex world.

Today those who seek to empower twenty-first-century students are charged with creating a pedagogy that prepares them to be global citizens. It is a time when students need to learn how to think, feel, and respond to the world in new ways. In this process, they must interface constantly with new information, diverse cultures and ideas, and challenges that require both critical and creative thinking. Helping students to cultivate their adaptive intelligence allows them to discover that *how* we know is at least as important as *what* we know. For Fuller, who was able to see the world from different points of view, this is fundamental for creative new insights and adaptive behaviors.

Fuller hoped for an age of "*omni-successful education and sustenance of all humanity.*" Recognizing the need for systems thinking, Bucky knew that what would be essential for success in this new era includes greater cognitive flexibility, comfort with the unfamiliar, appreciation of diverse perspectives, the ability to hold multiple points of view simultaneously, creative problem solving, and a capacity for discernment that relies on both intellect and intuition.

Each of these skills requires an appreciation for how worldviews—both our own and those of the people we are interacting with—shape our thoughts and actions. For educators in the twenty-first century, there is a growing awareness that students will interact daily with people who hold different perspectives, whose life circumstances and personal experiences inform their worldview in profoundly different ways. As students come into relationship with their social and physical environments, they develop habitual ways of viewing themselves and the world around them.

Human perceptions are filtered by the ways we view the world. Our worldviews influence every aspect of how we understand and interact with the world around us. Worldviews profoundly impact individual and shared goals and desires, shaping perceptions, motivations, and values—both consciously and unconsciously.

Worldviews inform human behavior in relationships and choreograph individual and social reactions and actions every moment of the day. They shape our habits of introspection, analysis, and communication, influencing the questions we ask, how we make meaning of our experiences, and ultimately the ways we live and learn.

In response to the need to understand worldviews and to encourage what Bucky called omni-successful education, the research and education team at the Institute of Noetic Sciences (IONS) created a curriculum on what we call Worldview Literacy (WVL). This involves the capacity to comprehend and communicate an understanding that information about the world around us is perceived and delivered through the filters of our personal and cultural worldviews.

It is the understanding that beliefs are embedded within individual and collective frames of reference and that other people hold different worldviews. Knowing that our worldviews or models of reality are largely unconscious and that jointly engaging in practices that raise our awareness of the beliefs and assumptions we hold can allow each of us to better navigate encounters with differing perspectives.

At IONS, we have created an experiential pedagogy designed to increase students' awareness of their worldviews. In this way, we open conversational spaces of exploration for students to explore their own beliefs and assumptions. In the process, students are encouraged to approach diverse worldviews with curiosity and wonder, in service to creating the deeper collective understanding and more effective sensemaking required to navigate life in the post-industrial, globally interconnected world.

As an experiential curriculum, WVL facilitates exploration of the pivotal role that worldview, perspective, or point of view plays in perception, information processing, and behavior. Students are encouraged to reflect and share their worldviews while gaining tools for understanding the worldviews of others. The goal is to use direct learning and guided self-reflection to help teachers and students cultivate meta-cognition, including awareness of worldviews, cognitive flexibility, and a capacity to hold conflicting information, as well as social and emotional skills involving perspective taking, connectedness, and pro-social attitudes and behaviors.

As Bucky noted, our children are coming *"into a more complex, more evolved universe than we can experience or than we can know. It is our privilege to see that new world through their eyes."* We've seen that the WVL project holds promise to equip students with the tools and skills they need to become global citizens in the twenty-first century.

We believe that when students develop an embodied sense that their own worldview is inextricably linked to their culture, region, religion, upbringing, environment, and personal experiences, and when they understand that everyone they meet is also *inside their own* worldview—then they will be better able to embrace the perspectives of others. This understanding can generate a greater connectedness, compassion and empathy—hopefully leading to more and better options for discovering and creating mutually desired futures.

Appreciating worldview is vital to the kind of complex thinking that Bucky advocated. Presaging the worldview literacy curriculum, Bucky noted that: *"The most important thing to teach your children is that the Sun does not rise and set. It is the Earth that revolves around the Sun. Then teach them the concepts of North, South, East, and West, and that they relate to where they happen to be on the planet's surface at that time. Everything else will follow."* In this way, our children can learn to navigate successfully in various environments, both social and geographic.

Acknowledging and mastering worldview provides a set of skills that can liberate students from the kind of mind-numbing education that Bucky was rightly critical of and that limits our ability to be multi-dimensional people. In this way, future generations can help guide the future of humanity, seeing the evolving world with creative new eyes.

MARILYN SCHLITZ Ph.D. is the President and CEO of the Institute of Noetic Sciences. She is a social anthropologist and a researcher who has worked in laboratory, clinical, and field-based settings. She is the founder of the Worldview Literacy Program.

⚠ BUCKMINSTER FULLER LIVED MUCH OF HIS LIFE IN "universe time," an often-unimaginable place I often describe as having one foot in "normal physical reality" and the other in the realm of infinite possibility. That's how he was able to stand in front of an audience for hours on end presenting his "thinking out loud" lectures.

He was not just sharing what he had learned from personal experience. He was also allowing the wisdom of the ages to flow through him with a childlike innocence that endeared him to audiences worldwide. And that wisdom of the elders that flowed through him also flows through those who appear to be the least experienced members of our crew onboard Spaceship Earth.

They have chosen to be born into this unique period so that they can teach us how to build Bucky's vision of "a world that works for everyone." They have the skills and the tools to transform our planet and our species from war, competition, fear, and scarcity to a genuine heaven on Earth where peace, cooperation, love, and abundance are the norm for all sentient beings.

To most adults, this would appear to be an insurmountable undertaking, but to the youth it seems natural and normal until they become indoctrinated into modern society and take the path of being "productive citizens." That term now generally refers to getting a "good education" and a "good job" so that one can earn a "good living."

Bucky often translated the term "earning a living" into "earning the right to live," and that shift is more evident today as the gap between the "haves" and the "have nots" continues to expand. In the world of infinite possibilities that our children and grandchildren can envision with their open hearts and minds, this disparity does not simply disappear—it never existed.

Their "more complex, more evolved universe" is one of interconnection, interdependence, integrity, intuition, imagination, and illumination. It is the reality that philosophers have dreamed of since the dawn of time, and it is here right now. All we need to do is learn from and respect the wisdom of the innocent, naïve, bright youth who have come here to our tiny, fragile Spaceship Earth to herald the dawn of a new era in which everyone shares the bounty of our abundant Universe and our precious planet.

## 5.4 "No human can prove, upon awakening, that they are the person who they think went to bed the night before or that anything they recollect is anything other than a convincing dream."

△ BUCKMINSTER FULLER WAS A PRAGMATIST WHO WAS ALWAYS seeking indisputable proof. However, after decades of examination and contemplation, he came to the same conclusion as many mystics—human existence is most likely a waking dream with great consistency, but most of us are unaware of that continuity. He often reminded his audiences that we can't even prove that the person who wakes up in the morning is the same person who went to sleep the night before. We can't guarantee a continuity of self at any time, even during the hours of sleep.

Even though most of us believe that our lives are a single ongoing experience, we can't absolutely prove that one moment is connected to the next or to the one before it. We also generally believe that we are our bodies or the work that we do, but a quick internal assessment proves that assertion to be untrue. Science has now proven that every cell of our bodies is completely replaced every seven years, so I am not currently the same physical being who wrote this sentence in 2011.

When each of us stops to consider what we are doing with our time, more and more people realize that their careers don't make a difference in the world or with other people. Some have only come to this realization when they found themselves replaced by a machine or another person. Just like the cells in our bodies, people in specific jobs are interchangeable. We can be replaced by another person who is not even on the same continent as the work that needs to be done. Some jobs can even be replaced by a machine.

This conundrum often leaves people pondering the challenging question, "who am I." Thankfully, Bucky spent many years considering this taxing question, and one of his most interesting answers he shared focuses on the concept he labeled "pattern integrity." Bucky said,

> "A pattern has an integrity independent of the medium by virtue of which it receives the information that it exists. Each individual is a unique pattern integrity. The pattern integrity of the human individual is evolutionary and not static."

In other words, we are each a distinctive "something" that is continuous, unique and evolving. If we were able to tap into our knowledge of that something, we might be able to argue the point that we are the same person who went to sleep the night before.

This is not completely true. Each of us is the same pattern integrity upon arising, but we have experienced and learned new things while we were asleep. Many people label these learnings and experiences as dreams, and those "dreams" change who we are and how we behave in physical reality. Thus, we are the same in some ways but very different in others.

It is believed that some people have achieved such "awakening" or "enlightenment" during a single lifetime. They knew their true identity deeply enough to maintain a continuous stream of consciousness through all aspects of their lives including sleep. Although he usually spoke humbly of himself as "an average healthy human," Bucky may have been one such person.

He was certainly clear about his mission here on Earth and he maintained his focus regardless of external circumstances. He also provided us with a great model of one human functioning as fully as possible in order to stay awake and contribute as fully as possible to the common good.

## 5.5 "If the success or failure of this planet, and of human beings, depended on how I am and what I do, how would I be? What would I do?"

*By Lisa Matheson*

GUEST COMMENTATOR

There are three posters hanging on the wall of our family room where everyone—family, neighbors, the posse of teenagers who are constantly hanging around, guests—gathers when they are in our home. They are simple posters, purchased through The Buckminster Fuller Institute. They didn't cost much, but have, for me, been worth their weight in inspirational gold. My favorite of the three, and the one that has become my personal mantra, is this quote by Bucky.

These posters went up on our wall in 2004, but it was much earlier that their words began to influence me. I was in my early 30s and was sure that I was missing something important, that there was more to life, not so much in terms of what I could have but in terms of what and how I could contribute. I was relatively successful at almost everything I had ever done; things had always come easy for me in sports, in the traditional path through business, in life. But, despite it all, there I was, smack in the middle of what I called my early-mid-life crisis.

It was then that I first "met" Bucky. I began studying with a group whose focus was on the work of Fuller and how it integrated with our core ideas around scarcity and abundance, education and science, money and business. Although by no means a mathematician at any level, I came to understand the basics of Synergetics, of synergistic results, of stability in universe.

At one of these group events in Tucson, Arizona, in the mid-1990s, I looked around me and saw an amazing collection of people from all over the planet. They were doctors, teachers, successful business people, mystics, athletes, and housewives, all fascinated by Bucky's work and where it might take them. There were only a few, however, who appeared to be integrating everything we were learning from Bucky into their day jobs.

Dr. Cheryl Clark, and my husband-to-be, Michael Healey, were both extremely interesting thinkers and educators—she in the criminal justice system in the State of New York and he in private business. What made them attractive to me at that moment, however, was that they were beyond Bucky enthusiasts. They were Bucky implementers, and I started to hang out with them.

Fast forward through a number of years of collaboration with both Cherie and Michael, as we sorted through how the science of Synergetics applied to social dynamic, and I had my "aha." I finally "got" what I would do if "success or failure of this planet, and of human beings, depended on how I am and what I do."

Social Synergetics™ is what emerged. It is the application of whole systems thinking to social organizations, structures, and relationships. It is a structure for solving any and all problems and doing more with less. It is a framework in which any content can work better, with less expenditure of resources.

The provenance of the term is this: SOCIAL = the interaction between people + SYNERGETICS = the study of systems. Put the two together and you have "the study of the systems of interaction"—how we relate to others one-on-one, in groups, on teams and in organizations.

From what I could see in the corporate world I interacted with each day, a new way of thinking, one that provided stability through the inevitable and omnipresent forces of chaos and change, was desperately needed.

But, I was not a traditional educator. I did not want to be at the front of the room. I was not patient enough to stand up and teach the way others could.

What I was good at, however, was what I call the "back-end." I am an idea synthesizer and a toolmaker. I can distill complex ideas down into visual and word patterns that are easy for audiences to digest. I can make sense of complexity. So, in collaboration with Cherie and Michael, I began to work on tools that would help people in organizations, whether they be corporate, social services, education, or not-for-profit learn to think synergetically.

Social Synergetics™ is still in its infancy, relative to Bucky's Synergetics. But it is fascinating to me. It is motivating, energizing, exciting. I feel completely aligned, every day, with it being my answer to Bucky's challenge.

Most of all, I experience a whole-being physical-emotional-mental-spiritual "charge" when people we work with get their "aha" around stable systems, around a model for thinking that helps them in a way that no other coaching or development work has. When they discover which of the points on the tetrahedron, or which vector has been the weak link in their project, in their management style, or in their business model, they make an exponential leap forward and know exactly what to correct. Their success is very exciting to me.

My husband wants to renovate the family room. He and I have become passionate collectors of local artisans, and he is starting to ask what piece will go up on the on the wall after the renovations.

Bucky's posters will. There is an energy to his words, to the quotes in these posters, that inspires me daily. When people come to our home for the first time and stand in front of them, I can see that some understand what Bucky is saying and others maybe not so much. But, I do believe that his words make everyone think just a little bit differently.

LISA MATHESON is the co-founder of Social Synergetics™and the President of its parent company, Water Communications, Inc. Lisa has over twenty-five years experience as a communications and training consultant, designing professional development programs and learning tools for corporate clients, as well as non-profit organizations. She is the co-author, along with Dr. Cheryl Clark of *DOING LIFE!® A Life Skills Program for Recovery from Addictions* and *SMART Choices! An Accelerated Life Skills Program.*

△ BUCKMINSTER FULLER CONSTANTLY ASKED HIMSELF THESE questions because he truly believed that the success or failure of our planet and all human beings depended on how he behaved and what actions he took. Those of us who take a similar position are finding that we can in fact make an enormous difference in the lives of others and in the evolution of our species. And a by-product of such a seemingly altruistic way of living is that the person being generous, kind, and grateful to all others receives benefits well beyond what she expects.

Bucky found this to be true time and time again, and he came to simply accept these supposed gifts from Universe that some would label "miracles" as an everyday occurrence. In other words, by constantly acting on behalf of all humankind, he was able to live a life filled with love, peace, joy, and abundance. That is not to say that he did not have many challenges and struggles, but he did not judge these aspects of his life as negative. Instead, he simply saw them as more opportunities to learn, and he was soon moving beyond one challenge to something even larger and more beneficial to everyone (including himself and his family).

Using trial and error, he quickly learned that he could course correct for anything that did not produce a "positive" result and that the larger the challenge he gracefully accepted, the less he had to sweat the small stuff. For example, when someone like Bucky (or you or me) takes on the challenge of housing or feeding all humankind, things like his relationship with friends or family or where he will live become minor issues. We don't have time to focus on such issues when we're saving the humans, and they seem to simply take care of themselves. In other words, the seemingly huge issues in our lives become minor irritations when we come from a larger perspective as Bucky did.

When acting "as if" the success or failure of human beings is your responsibility, you naturally act more responsibly in all areas of your life because you know that your every action (whether witnessed by another or not) does matter and will make a difference. This is the law of cause and effect, and the Buddhists call it karma.

Working from this perspective, we realize that everything matters—be it picking up a random piece of litter, saying thank you to an angry clerk, or giving an inspirational talk to thousands of people. It all counts toward the success or failure of what Bucky described as *"humanity's final examination"* to answer the cosmic question, *"are humans a worthwhile to Universe experiment?"*

This is something that we each need to answer for ourselves, and things work best if we consider the question and respond remembering that the success or failure of this planet and human beings does in fact depend on how we are and what we do, right here and right now.

## 5.6 "God is a verb."

*By David Spangler*

GUEST COMMENTATOR

Like many geniuses, Bucky Fuller has left us an extensive and wonderful treasure of quotable statements. Picking one is like being asked to choose one's favorite food amongst a buffet of culinary delights where each item has its own unique flavor and attraction. It's a task both difficult and enjoyable at the same time.

In my case, I chose Bucky's statement that God is a verb in part because it succinctly expresses my own experience of sacredness and in part because if we could all appreciate this one simple fact, it would go a very long way to diminishing if not outright removing many of the religious conflicts that in one way or another continue to divide our world. And, I believe, it would enable us to recognize God in this world, seeing sacredness not as a transcendent reality somewhere "out there," but as something inherent in the processes that enable life to exist. We would then be in a much better position to recognize our own sacredness.

Nearly a hundred years ago, Dion Fortune, an Englishwoman who was one of the leading teachers in the Western esoteric tradition, said something similar. She said, "God is pressure." This also captures the same sentiment to which Bucky was directing our attention, that the sacred is an activity, not a thing. Further, it is not simply the activity of bringing universes into being but the activity that allows all other activities to proceed. When I walk, when I talk, when I engage the world in creative activity, all this is possible because of the activity that is God. Sacredness is a doing.

In spiritual circles one can often find a great deal of effort expended in discovering what it means to be sacred or in becoming sacred. But this assumes that sacredness is a condition that we can discover and enter. When we understand, as Bucky did, that it is a doing more than a being, then our attention goes to how we are acting. God appears in our actions, in our behavior, in our doings. The sacredness manifests in how we engage the world. We can sit and "be" all day long, and the world will not be the better for it. But when any one of us engages in the process of sacredness—in the activity of a comforting smile, a friendly word, a kindly action, a loving goodwill, a compassionate helpfulness—then the world can feel the presence of the sacred in its midst.

My own work focuses on the development and practice of an incarnational spirituality, a spirituality that celebrates our human personhood, our connections to the Earth, and our physical lives. Such spirituality deeply honors the perspective of sacredness that Bucky articulated in this simple quote. Its purpose, too, is to make it possible for us to look at ourselves and to look out at our world and to see in all the processes that surround us the living presence of sacredness. Even more importantly, it presents an image of a sacredness with which we can partner in accessible ways.

For instance, if I feel I have to be a fireman and wear a fireman's uniform to fight a fire, then I may stand by uselessly while my house burns, waiting for real firemen to arrive. But if I understand that being a firefighter is a verb, then I can grab my garden hose and fight the fire even though I don't have the uniform or the title.

If God is a noun, how can I be sacred? How can I act in sacred ways in my world? How do I become a god myself? But if God is a verb, then any of us can "do" Godliness and in so doing, make the sacred visible.

I believe our world needs—even requires—us to do sacredness if we are to have a positive future. Bucky understood this, I'm sure, and it informed his creative ability to open new doors and unveil new possibilities throughout his life. The world awaits the rest of us to do the same. After all, we are verbs as well.

DAVID SPANGLER is a spiritual teacher, a former co-director of the Findhorn Foundation in northern Scotland, a Lindisfarne Fellow, and one of the founders of the Lorian Association. He is the author of *Apprenticed to Spirit*, *Blessing: The Art and the Practice*, *Facing the Future*, and *Subtle Worlds: An Explorer's Field Notes*. He teaches classes online and publishes a monthly email essay, David's Desk, and a quarterly print journal, *View From the Borderlands*. For information about his work, his books, and his publications, please go to www.lorian.org.

FOR BUCKMINSTER FULLER, GOD WAS MUCH MORE THAN a noun—a being that created and supported human life. Although he believed that organized religions were detrimental to the evolution of humankind and the success of all people, Bucky was a very spiritual person, and his concept of what we label God is as an active inner and outer intelligence that is well beyond words.

God is all pervasive and undefinable. It is within the physical being we called Bucky Fuller and is within each of us. It connects us with one another and explodes in the bounty we experience as life on Earth. It cannot be created or destroyed, is infinite and therefore is something most of us cannot wrap our minds around.

Bucky often attempted to understand and explain the phenomenon we call God, but he was rarely able to do so in his writing. His live presentations were far more successful as he was able to source and transmit an essence that some have labeled as Source, Greater Intelligence, Great Spirit, Higher Power, or Christ Consciousness.

Many people would leave his marathon "thinking out loud" lectures after four or five hours saying, "That was the most amazing experience of my life, and I have no idea what he said." Years or decades later, individuals recount how much just being with Bucky had transformed their lives.

As Barbara Marx Hubbard recounts in her Chapter 1 commentary for this book, Bucky once told her that "he wanted to support my writings. When I asked for an endorsement he wrote:

*'There is no doubt in my mind that Barbara Hubbard, who founded the Committee for the Future and helped introduce the concept of futurism to society, is the best informed human now alive regarding that movement and the foresights that it has produced.'*

What he meant, I believe, is that both of us had a Christ experience that revealed to us the evolution of humanity from a self-centered, creature human to a whole-centered, spirit-centered universal human, at one with the Processes of Creation and Nature."

This "Christ experience" is the essence of conscious awakening that Bucky and other spiritually adept masters transmit to us. It is given to them, and they are empowered to pass it on to each of us in support of our waking up and moving closer to complete enlightenment. The verb God within each of us will continue to be active and activate that awakening within others until we as a species take that next evolutionary leap into a form we cannot yet imagine. Then, like Bucky himself, it will continue working to bring us to the next level of consciousness.

# Consciously Empowering Others

## Chapter Summary

**5.7** **"I don't know anything about the subconscious, and I have no right to deal in it. And so I have to deal in the conscious, and there has to be an attempt to communicate the experiences."**

BUCKMINSTER FULLER OFTEN SPOKE AND WROTE ABOUT realizing how little he knew, especially as he grew older. He may have claimed that he knew nothing about the subconscious, but he was clearly able to access a consciousness that is more comprehensive and potent than most of us intentionally know or experience.

That awareness—which he shared in his writings, lectures, and inventions—models something infinite and beyond the scope of the human imagination. Still, like Bucky, each of us attempts to imagine it and give words to our imaginings. We write songs about it. We make art to share our experience of it. And if we get overenthusiastic about it, we try to convert others to our way of seeing and understanding this subconscious/unconscious.

Others usually resist such proselytizing, and Bucky learned that early in life. After some years of attempting that strategy, Bucky realized that it would never work, and he stopped trying to convince or change anyone. Even though he had had many profound experiences connecting to Source and had been given the ability to communicate a Source experience to others, he allowed people to find him rather than advertising his gifts.

Realizing that he could not change people and giving up even trying is the mark of a genuine spiritually awake individual. They are humble yet willing to fully give to those who show up. They help people to awake to the fact that they are awake within a waking dream that we call reality. And they never seek fame, wealth, glory, or power for themselves.

Life for them is what Helen Keller famously labeled "a grand adventure" when she said, "Life is either a grand adventure or nothing." People like Bucky and Keller are the true consciousness teachers and leaders, and they have been empowering others to follow in their footsteps for generations. Bucky received one such empowerment from Albert Einstein, and he gave many empowerments to those who wanted to learn and grow with him.

It was never his intention to be put on a pedestal or even to be recognized as a wise expert. He wanted to simply be who he labeled himself—"a comprehensive, anticipatory design scientist" and an "average, little man." May those of us who are similarly blessed with the conscious recognition of who we are continue to mindfully and generously share this awakened heart with all who seek it.

## CHAPTER 6

# UNITY

## IT'S EVERYBODY OR NOBODY

## 6.1 "Either war is obsolete or men are."

THIS SIMPLE STATEMENT IS MORE TRUE AND RELEVANT THAN when Buckminster Fuller made it in the 1970s. Either we shift our resources from weaponry to livingry, or we will soon be an extinct species. War became obsolete, just as Bucky predicted, in approximately 1976—when we reached a point of doing so much more with so many fewer resources that there was (and still are) enough resources to support all life on Earth. Unfortunately, most people do not know this, and wars continue to destroy lives and cultures.

Yet "we the people" continue to follow the dictates of a few greedy, self-serving individuals who have been granted the mantle of power by a system that passes it down from one elite generation to the next and bypasses true democracy or freedom. These few people in power would have us continue to believe that we live in a universe of great scarcity on a planet of scarcity where we must battle one another for limited resources.

They want us to believe in the reality that it's a "you *or* me" world rather than waking up to the fact that we live in an abundant Universe on a "you *and* me" planet. They adhere to the "divide and conquer" policy that kept their ancestors in power and continues to keep them dominating human existence at the expense of a successful life for all. And they want to keep "we the people" isolated from one another because they know "to keep conquered, keep divided."

In other words, they want us to keep thinking that we are "rugged individuals" able to make it on our own and fighting for our piece of the pie when this is exactly what is leading to the end of human existence. Fear, war, and competition will result in the extinction of the human species, and this reality is much closer than any of us realize. We could become extinct before this book is published. If you're reading this, we still have time.

Buckminster Fuller warned of this reality for the majority of his "56 Year Experiment" from 1927 until his death in 1983, and his final book focuses on that issue. In that diminutive 1983 book, *Grunch of Giants*, he writes about the "Gross Universe Cash Heist," which he calls the GRUNCH. He metaphorically describes the Grunch as *an army of invisible giants, one thousand miles tall, with their arms interlinked, girding the planet Earth.*

Although this frightening vision is relegated to science fiction movies, the institution Bucky was describing does exist and is strangling humankind to death. The Grunch are the industrial leaders and politicians who gain enormous profit from war and exploitation of our beloved Mother Earth while fifty thousand people die every day of starvation.

These are the men (yes, they are mostly men) who wear suits or uniforms with lots of medals to impress us when they are recorded telling us their version of the state of our planet and our individual nations. And what they tell us as hard fact is that there isn't enough. They warn us that we have to be wary of "foreigners" and those who would take our share of the pie. And they continue to heist all our resources (often in the form of cash) and hoard their ill-gotten gain because they believe that having a personal storehouse of resources will somehow protect them from a societal collapse.

In their antiquated thinking, they believe that having lots of money will make a difference when there is no clean air to breathe, fresh water to drink, or uncontaminated food to eat. They think that a few people can survive a global nuclear war or other man-made catastrophe when nothing could be farther from the truth. Whether we like it or not, we're all in this together. We have no backup planet, and there are no spaceships waiting to take the chosen few to live somewhere other than Earth.

The Grunch is real, and it is the scourge of every sentient being on Earth. It can also be eliminated as can its propensity to create more war so that "we the people" can finally have our heaven on Earth, or as Bucky described, our natural state—*a world that works for everyone.*" We can choose to cooperate and end war, or we humans will go the way of the dinosaurs—yet another failed experiment on this planet. It's up to each of us to decide our fate as well as the fate of potential future generations and all sentient beings on our tiny, fragile Spaceship Earth.

## 6.2 "I am a passenger on the Spaceship Earth."

*By Kevin J. Todeschi*

GUEST COMMENTATOR

As a longtime resource for Edgar Cayce information, I would have to say that my favorite Bucky quote is the one-line in which he states, *"I am a passenger on the Spaceship Earth."* Although it can certainly be seen by some as a simple statement, for me it embodies a rich array of possibilities and truths about the nature of humankind and the interconnectedness of the entire human family.

The quote does not state that I am piloting the vessel, nor does it imply that my seat is in First Class whereas someone else may have been relegated to Coach—I am a passenger, no more important (and no less important) than any other passenger. There is also a deep implication about our connectedness on this journey—although we may not always be clear about our ultimate destination or perhaps even how long we will be able to participate in the ride, what is clear is that we are traveling together. There is a oneness to our collective voyage that is too often overlooked when we choose to focus only upon the importance and value of our individual life's journey.

It's clear that Bucky was aware of the importance of the Whole as well as the importance of each individual piece. Unfortunately at the present time in too many places on our Spaceship Earth, it appears as though we are focusing upon the importance of one of these extremes while disregarding (or even attacking) the other. At the very least this has contributed to personal imbalance, selfishness, self-importance, and neglect, on the one hand, while it has squashed freedom, personal democracy, the capacity for change, and individual liberties on the other.

In the West, we take pride in our individual freedoms but too often downplay our need to work together as a society, as a country, as collective citizens of the planet. Too often the idea seems to be, "I am more important than you." In other places on Spaceship Earth, we often champion what collective society has done to improve the human condition or address human needs but we shy away from individuals who are different, unique, or have a voice apart from the group, society, or country to which we belong. Too often, the idea seems to be, "If you are different, you don't belong." Neither of these extremes will work to solve the problems we must address as a human family.

The world has changed dramatically in the past couple of generations. In all likelihood, our grandparents never imagined a world in which anything occurring on one side of the globe would so quickly affect the other. We are connected because of food, energy, health, weather, catastrophe—the list goes on and on. To solve the issues facing humankind—issues related to food shortages, water problems, energy costs and availability, etc.—will require all of us to work together for the good of the whole. We will need to understand and work with our interconnection perhaps in ways that we have yet dared to imagine. For though we may be separated by religion, or language, or culture, or distance, or politics, we are all passengers on the Spaceship Earth; we are all travelers along the way.

KEVIN J. TODESCHI is the Executive Director and CEO of Edgar Cayce's A.R.E. and Atlantic University. As both student and teacher of the Edgar Cayce material for more than thirty years, he has lectured on five continents to thousands of individuals. A prolific writer, he is the author of hundreds of articles and more than twenty books, including *Edgar Cayce on the Akashic Records*; *Dreams, Images and Symbols*; and several novels, including *The Reincarnation of Clara* and *The Rest of the Noah Story*. More information about the Cayce work is available at EdgarCayce.org and AtlanticUniv.edu.

△ BUCKMINSTER FULLER MADE THIS SEEMINGLY STRAIGHT-forward unassuming statement often to remind himself and his audiences where we are and what we are doing. We are passengers, guests on a beautifully designed, sustainable Spaceship we call Earth. If we fulfill our function as "local information gatherers and problem solvers in support of an eternally regenerative Universe," we might also consider our-selves to be crewmembers on board Spaceship Earth.

However, we must also recall another seemingly straight-forward unassuming statement Bucky also shared with his audiences when he would remind them, *"The most important thing about me is that I am an average man."*

We are all average people traveling as guest passengers on Spaceship Earth, and we would do well to recall those facts on a daily basis. Then, we could consider our actions and ask ourselves if:

- we are well-behaved, polite, respectful passengers;
- we believe ourselves to be better or more deserving than other passengers;
- we are maintaining our vessel so that others and future generations can enjoy its bounty and beauty;
- we can change our behavior and act in ways that honor and respect both our living Spaceship and our fellow passengers;
- we are modeling appropriate behavior for our children and future generations.

These are all questions that Bucky posed to his fellow pas-sengers in his diminutive 1968 book, *Operating Manual for Spaceship Earth,* and to audiences throughout his life. They are not complicated or beyond the reach of any person when we recognize our place here on Earth. We are just people doing the best that we can, but we are also capable of much more.

Buckminster Fuller devoted fifty-six years of his life to a personal experiment determining and documenting what one individual could achieve that could not be accomplished by any government, corporation, religion, or other institution no matter how large or powerful. The result was a life that is generally acknowledged as well lived and models genuine contribution to all passengers on board Spaceship Earth.

And being *"an average man,"* he often reminds us that each of us can accomplish much more than he was able to achieve. All we need to do is consider what type of passenger/guest we want to be and act from that context.

We don't need to make huge efforts to change the world or even change those people we believe are not behaving properly. We don't need to try reforming the system, especially in light of the fact that the system is broken beyond repair. And we certainly don't need to be martyrs sacrificing our well-being for the sake of others.

In an abundant world that works for everyone, we simply need to realize:

- Who and where we are—passengers on Spaceship Earth traveling through the galaxy at about 66,660 miles per hour and spinning at approximately 1,037 to 180 miles per hour (depending on where you are located).
- What we are here to do—to be local information gatherers and problem solvers.

Then, all we need to do is look at our daily actions and see if we are in fact gathering information and solving problems in ways that support sustainable abundance while we zoom through the Universe spinning like a top. Welcome to the E Ticket ride you've been on since the day you were born.

**6.3** "It is now highly feasible to take care of everybody on Earth at a higher standard of living than any have ever known. It no longer has to be you or me. Selfishness is unnecessary. War is obsolete. It is a matter of converting the high technology from weaponry to livingry."

*By Lynne Twist*

GUEST COMMENTATOR

Bucky was called the Grandfather of the Future. He was one of my teachers in the latter part of his life. He was such a beautiful being, I just wanted to be around him.

I was a volunteer usher in 1976 for one of the eighty "integrity talks" he gave around the world upon turning eighty. Bucky always stood behind a table with models that explained his theory about the construction of the Universe. I didn't understand any of what he said. *And* I remember everything else about that moment, a moment of absolute clarity.

In a talk about the intellectual integrity of the Universe, he stepped out from behind the table and said, *"Now I'm going to say the most important thing I have ever said or ever will say."* I thought, "I *am* going to understand this!" So I tuned my antennae and sat on the edge of my seat.

Bucky said that humanity had recently passed a very critical threshold. When he said "recently," he meant sometime during the last fifty or 100 years. He then went on to say this threshold changes everything. The threshold is this: humanity is now on a path of doing so much more with so much less, which is at the heart of human innovation and creativity. And we now clearly live in a world where there is enough for everyone, everywhere, to have a healthy and productive life. We now live in a world of sufficient resources, a world of "enough."

Bucky said this means we have moved from a "you *or* me world" (because there's not enough for both of us—the mindset of scarcity), to the possibility of a "you *and* me world," where we both can make it at no one's expense. That new possibility changes everything. He speculated that perhaps it has always been that way, and it's clearly true now.

In that moment I started crying and had an epiphany, a revelation that was almost more than I could handle. I don't know if I understood the words he said at that time, but I totally "got it."

Bucky said it would take fifty years for humanity to come to terms with this new paradigm because all human institutions are rooted in the belief of a "you-*or*-me" paradigm—for example, education, governance, the economy, and even religion. He predicted it would take fifty years to rethink human institutions in a context of "you-*and*-me," a context of "enough" or sufficiency.

I started to see the world in a completely different way. There was enough of everything. All needs were being met by the Universe; maybe not all "wants," but all "needs." In other words, the Universe had within it an ethic of sufficiency.

It was after that that I started working at The Hunger Project, in which Bucky had an important part of founding. What I saw over twenty years in places like India, Bangladesh, Sub-Saharan Africa, Ethiopia, and Senegal—all places of hunger and poverty—was the "enoughness" of human beings, and that people *could* cope with and meet their needs. I also saw the beauty of who we are as a human family.

This revelation changed everything for me. I became a person living in the context of sufficiency, which I call the "radical, surprising truth." This understanding is waiting for us to see it when we clear away the mindset of scarcity. What we see is that sufficiency or "enough" in a "you-*and*-me world" is consistent with the capacity to love and care for each other, the taproot of true "security" and community.

Out of that whole new understanding of the world grew the *Principle of Sufficiency*: If you let go of trying to get more of what you don't really need, which is what we're trained to want more of, it frees up oceans of energy to pay attention to what you already have. When you nurture and nourish what you do have and begin to make a difference with it, it expands before your very eyes. In other words, what you appreciate appreciates. This is true prosperity.

Sufficiency is different than "abundance," which is the flip side of scarcity in our consumer culture. It's not an amount. Rather, it's a way of seeing, a way of being. When you lift the veil, you see exactly what you need. When you build from "enough," true abundance is yours.

You can see the power of appreciation and "enough" operating in your own life—for example, when you appreciate people, they flourish. When you can appreciate the darkness of a difficult period of your life, you can see how you were met by the Universe to give you exactly what you needed to find your own way and become self-sufficient.

I wrote *The Soul of Money: Transforming Your Relationship with Money and Life* and founded the Soul of Money Institute to support people in finding power, freedom, and peace in their relationships with money, each other, and themselves.

I am forever indebted to Buckminster Fuller for illuminating this path.

LYNNE TWIST is Co-Founder and Board Member of The Pachamama Alliance. She is author of the award-winning book, *The Soul of Money*, and President and Founder of the Soul of Money Institute. Lynne is also a global activist, fundraiser, speaker, and consultant, who has devoted her life to service in support of eradicating hunger and poverty, empowerment of women and girls, global sustainability and security, human rights, economic integrity and spiritual authenticity. She is also principal consultant to the Nobel Women's (Peace) Initiative. For more information about her work, please visit www.soulofmoney.org and www.pachamama.org.

△ BUCKMINSTER FULLER'S PHILOSOPHY OF SURVIVING AND thriving on Earth can be boiled down to the fact that shifting from weaponry to livingry is the solution to all humankind's current physical problems. It's just that simple. All we need to do is shift approximately 40% of our global military budget from things that take life (weaponry) to things that support life (livingry—food, education, transportation, housing, etc.).

In the 1930s Bucky manually made an inventory of all the world's resources and correctly calculated that we were doing so much more with fewer resources that we would reach a point when there would be enough to take care of everybody at a higher standard of living than anyone has known. He also determined that the paradigm shift would take place in 1976.

It has now been proven that in 1976 we reached the point where there was enough food on Earth to feed everyone. Still, tens of thousands of people continue to die of starvation every day while a huge percentage of our resources remain focused on weaponry rather than livingry. This is true in every area of human existence.

We live on an abundant planet, and each of us can have all that we want and need without another person going without. In that plentiful environment, war and the politics of competition are obsolete. We simply need to wake up to the reality of cooperation in which we are all rich beyond our wildest dreams. Then, everyone can focus their time and energy on doing the things that they most love and contribute their talents to supporting others.

This may sound like some pie-in-the-sky vision of a future that will never happen, but it is the context of an abundant world Bucky proved possible and championed most of his life. It's also the perspective that each of us must embrace if we are to survive and thrive as individuals and as a species.

As Bucky so often reminded his audiences, we're six billion billionaires who have yet to recognize our true nature as billionaires. We're the beneficiaries of the suffering, mistakes, and learning of past generations, and we have it all within our grasp if we're willing to reach out, take control of our destiny and shift from competition to cooperation, from division to unity, and from weaponry to livingry.

This is how we "take care of everybody on Earth at a higher standard of living than any have ever known." For most people, this appears to be something well beyond the possible. Bucky would, however, continually remind his audiences that when he was young the wealthiest king or businessman did not have refrigeration, wireless communication, telephone service, or the ability to fly anywhere in the world. Thus, many people today live at a higher standard of living than that king or corporate magnate.

Today we only need look in our pocket or purse to notice our much higher standard of living. Only twenty short years ago nobody had a small, inexpensive cell phone, much less a smart phone that was also a camera, news source, texting tool, and music player. That single innovation led to many of us enjoying a higher standard of living than our parents and their peers could even imagine, and the number of other such innovations is huge.

Bucky was one of the first (if not the first) people to recognize this enormous yet subtle shift in our environment and ability to care for one another. Now, however, it's time for each of us to recognize this new paradigm and quickly adapt it into our daily lives. This is not a question of "if," but rather an issue of "when." We must act quickly to educate ourselves and our fellow crewmembers aboard our tiny, abundant Spaceship Earth if we are to survive and thrive as evolving beings acting as stewards for all life.

**6.4** **"There are no 'good' or 'bad' people, no matter how offensive or eccentric to society they may seem. You and I didn't design people. God designed people. What I am trying to do is to discover why God included humans in Universe."**

*By Jamal Rahman*

GUEST COMMENTATOR

With these words, Buckminster Fuller opens a whole new vision for the way we see ourselves and each other. Rather than focusing on the conditioned personality, which may indeed be good or bad, we need to realize that all people are manifestations of the Creator, possessed of divine potential at the core of their being. In Islam we say that humans are essentially *fitra*, or good, but capable of making bad choices that mask our divine essence. A similar teaching is at the heart of every spiritual tradition.

Thus, in our dealing with others, it is important to distinguish between behavior and being. A person's behavior may sometimes be evil, but the *person* cannot be evil. Each person's essential being is sacred, filled with the spirit of the divine, whether we call that divinity God, Allah, Adonai, Krishna, Buddha, Jesus, or simply The Universe.

This is not to say we should simply accept someone's bad behavior and let ourselves be run over. In the words of the sixteenth-century Sufi sage Kabir, "Do what is right. Protect yourself. Don't allow yourself to be abused. But, please, do not keep the other person's being out of your heart." Simply remembering the vital distinction between behavior and essence will influence how we speak and act toward others, and this, the sages say, has the power to shift heaven and earth.

In our current political climate of Islamophobia, I am acutely aware of the dangers of not making this distinction. Too easily we label others as "evil" simply because of their religion, their color, or their manner of dress. Often we condemn an entire people because of the actions of a few misguided persons. When we lapse into this unconscious mode, it affects our behavior and produces hurtful consequences

We claim a higher moral ground and refuse to dialogue. We justify and rely on the use of force. We delude ourselves away from the truth and avoid the real work. Alexander Solzhenitsyn, the Russian dissident, sums it up succinctly: "If only there were evil people somewhere insidiously committing evil deeds and it were necessary only to separate them from the rest of us and destroy them. But the line dividing good and evil cuts through the heart of every human being. And who is willing to destroy a piece of his own heart?"

The real work, then, is to focus on our own hearts, gradually learning to let go of our conditioned reactions and connecting to our divine nature. Every saint has a past—a life once lived without being conscious of sacred potential. And every sinner has a future—the possibility of joyful connection between personality and essence.

The surest way to achieve that connection is to continuously and compassionately witness our speech and actions as we go about our daily lives, trying as often as possible to act from divine essence rather than from ego. The ego is a necessary instrument of the soul and cannot be destroyed, but it can be transformed by continuous efforts to align it with the divine attributes of the soul. "Marry your soul," said the thirteenth-century Muslim Sufi sage Rumi. "That wedding is the Way."

JAMAL RAHMAN is co-founder and Muslim Sufi Imam at Interfaith Community Church and adjunct faculty at Seattle University. His books include *The Fragrance of Faith, Out Of Darkness Into Light, Getting to the Heart Of Interfaith* and *Religion Gone Astray*. He travels nationally and internationally with a Rabbi and Christian Pastor in a unique emerging ministry—affectionately known as the Interfaith Amigos—spreading a message of inclusive spirituality. Learn more about Jamal and his work at his websites www.jamalrahman.com and www.interfaithamigos.com.

△ BUCKMINSTER FULLER'S ONGOING DECLARATION OF NON-attachment to judgments—especially regarding other people—is the perspective espoused by many spiritual masters over the years. Thus, some consider him to be an enlightened, spiritual being.

Rather than judging Bucky for this and other seemingly controversial statements, it might be better to consider adopting his global perspective and his ability to include all beings in his work. He realized that he could not know which person, animal, or sentient being was more important than another, so he treated everyone equally.

He often reminded his audiences that although people sometimes believe they are all powerful, we humans are not *"running the show"* anywhere on Earth. Because of that fact, Bucky (and many wise individuals) took time to discover why we are here and work to further the concept of a world that works for everyone. That's what Bucky did, and it provided him with both satisfaction and a life well lived.

When asked why humans are here on Earth, Bucky was quick to recount that the purpose of humans on Earth was as local information gatherers and problem solvers in support of the integrity of a fully regenerative and sustainable Universe. Still, he constantly wondered why God included us in the scheme of things since we don't seem to be doing a great job of fulfilling our mission as stewards on Earth.

Bucky was clear that his individual mission was to work on behalf of as many sentient beings as possible. When writing about that mission, he said,

*"The challenge is to make the world work for 100% of humanity in the shortest possible time, with spontaneous cooperation and without ecological damage or disadvantage of anyone."*

With that statement he reiterated the fact that inclusion was an enormous part of his work, operating strategy, and life disciplines. He did his best not to judge others and to include as many people as possible in all his endeavors. He understood the power of synergy (a chemical term that he popularized), and he worked to have synergy, integrity, and teamwork in all his initiatives.

This was not an easy task at a time when most people believed that people whose appearance, lifestyle, or country were different were dangerous "foreigners" to be avoided and defended against. That same culture of fear and exclusion continues today, but there are an increasing number of people who recognized, as Bucky did, that we are a single interconnected and interdependent species.

During the period when Bucky was championing this unity and oneness, most people were building larger walls and borders to protect themselves. Now that many of us are aware of the fact that we are one species surviving on a tiny, fragile lifeboat we call Earth, Bucky's insights become even more significant.

We know that we do not control Universe, Earth, or even the places we call home. Natural conditions (often labeled disasters) demonstrate the true power of the environments we humans have attempted to control and exploit. The singular cosmic question that Bucky was always reminding us of is, *"Are humans a worthwhile to Universe invention?"*

Each of us has to answer this inquiry for ourselves. Are we doing what needs to be done on behalf of all humankind? Are we including as many others as possible in our efforts and solutions? And are we operating with personal integrity in all aspects of our lives?

These are some of the questions Bucky continually asked of himself. Following the discipline of remaining true to his personal integrity and mission, he was able to make an enormous contribution to humankind and all life on Earth. He did not judge anyone as not worthy of his support, and all sentient beings benefited greatly from his contributions. Hopefully, many of us can continue on the path of contribution, cooperation, equanimity, integrity, and inclusion that he so elegantly mapped out for us.

**6.5** "Think about a stadium of seventy-five thousand people. It's really quite a large crowd, isn't it? Every day seventy-five thousand people die of starvation despite the fact that we have plenty of food for everyone. Our distribution systems, our nations, all the different kind of separateness blocks the whole thing. Just think of it. Simply because we're badly organized, we're not taking care of it."

THIS WAS AN ASTONISHING STATEMENT WHEN BUCKMINSTER Fuller first made it in 1976, and is even more staggering today because little has changed. Our human population has increased and so has the number of people who die of starvation every day even though there still is plenty of food to provide for everyone.

And the blockage to feeding those people and supporting all life on Earth remains exactly what it was decades ago—a system that supports greed, competition, and war. We live in an era that must be focused on abundance for all, cooperation, and the rapid emergence of a peaceful, sustainable environment for all beings.

This is not some pie-in-the-sky dream of a visionary idealist. It's not the ideal of a delusional radical. And it's not the dictum of a man seeking followers.

It is the vision of Bucky Fuller, a self-described "comprehensive, anticipatory design scientist" who devoted the majority of his life to a disciplined study of humankind and to finding and sharing as many possibilities for our success as possible. He did this primarily by seeking solutions to problems that he believed would emerge fifty to one hundred years in the future. He also realized that the main challenge facing our species would be to recognize the obsolescence of our fear- and scarcity-based systems and to quickly replace them with unique new structures that have integrity and serve the needs of all people.

In this simple anecdote regarding food distribution and seeming scarcity, Bucky shared a microcosm of the macrocosmic issue—the often unnoticed paradigm shift from a global culture in which there was not enough to support all life to a culture of abundance. We now know that we have ample resources to sustainably support all life if we remove blockages including sovereign nations, corporations, religions, and other entities that promote separateness rather than interconnectedness and interdependence.

We are a single species living on a tiny, fragile Spaceship Earth, and we are fortunate that in 1936 Bucky began his research to determine when we would shift from a culture without adequate resources to support everyone to a society in which there is enough to go around. He determined that date to be 1976, and we now know that his prediction was correct when applied to food.

Modern science and economics has proven that in 1976 humankind did in fact reach a point of having enough food to support everyone. The individuals who did not seem to "get the memo" about this shift were (and still are) those we usually refer to as "leaders." They are the CEOs of corporations, religious leaders of all faiths, presidents of countries, and others who benefit from controlling masses of people. Almost all of these "leaders" preach the gospel of "divide and conquer and to keep conquered, keep divided," and it is that divisive strategy that Bucky contended has kept us from bringing *"a world that works for everyone"* to fruition.

This small group of "leaders" use the media and other propaganda tools to convince average citizens that there is not enough to go around and that only "leaders" have the solutions to all our problems. Unfortunately, their solutions focus on their nation, religion, corporation, etc., being better than any other and thus being entitled to a greater share of what they tell us are limited resources.

The result of the divisiveness that they foist on "we the people" is our current situation in which tens of thousands die needlessly of starvation every day while we spend billions of dollars every day on a worthless military system. In an abundant Universe and global culture, there is no need to fight for a share of the pie. Since there is now plenty to go around, all we need to do is recognize this fact and begin the process of making sure that each and every human and living being on our tiny Spaceship Earth is fed and fully supported.

**6.6** "If humanity does not opt for integrity we are through completely. It is absolutely touch and go. Each one of us could make the difference."

*By Dr. David Gruder*

GUEST COMMENTATOR

Integrity, happiness, and profits got a divorce during the twentieth century. The installation in the 1950s of the faulty happiness formula known as "The American Dream" sealed the deal on this divorce.

The happiness formula for democracy-centered republics was originally described in the United States Declaration of Independence and Constitution. Its highest value was ensuring that society was structured around the right of its people to the pursuit of happiness. The formula for the pursuit of happiness revolved around citizens and their government navigating a seeming paradox between preserving personal freedom and promoting the common good.

The English language doesn't have a word for the equal attention to freedom and responsibility. I therefore coined one in my six-award-winning book, *The New IQ: How Integrity Intelligence Serves You, Your Relationships and Our World.* That word is "Freesponsibility." The U.S. Constitution's framers understood that living a Freesponsible life was a citizen's and government's fundamental act of integrity in order to make the pursuit of happiness possible.

Prosperity that grew from this version of the pursuit of happiness was the essence of socially responsible profitability. Napoleon Hill, in his classic book *Think and Grow Rich* succinctly captured the essence of this. The secret to the pursuit of happiness in the world of business is creating one that "works with and for the public, yet is still profitable."

Over my decades of experience as a clinical and organizational psychologist, I have found that there is a universal and timeless formula for both integrity and happiness. This formula contains three elements: authenticity, connection, and impact. Authenticity is being who you truly are. Connection is feeling bonded with others. Impact is having a positive influence on the world around you. I have seen over and over that those who experience sustainable happiness experience all three of these dimensions in their daily life, not just one or two of them.

Here's the secret that ties integrity and happiness together: authenticity is impossible without self-integrity, connection is impossible without relationship integrity, and positive impact is impossible without collective/societal integrity. I refer to this as 3D Happiness and Integrity.

This is why happiness and integrity are inseparable. Any attempt to separate them results, by definition, in a faulty happiness formula and integrity deficits. Unfortunately, this is precisely what happened during the 1950s, much to Bucky Fuller's chagrin I am quite sure.

The faulty 1950s version of the American Dream was built around a very different formula than the original version of the American Dream. The 1950s version re-defined the pursuit of happiness as excessive consumerism. It replaced preserving individual freedom with the expectation that we conform to the delusional notion that excessive consumerism is the secret to happiness. And it replaced promoting the common good with the requirement that we overwork in order to make the money we need to succeed at excessive consumerism.

Family breadwinners were led to believe that if they loved their spouse, children, the business they worked for, and society in general, they should sacrifice their passion and life balance and do whatever they needed to do to bring home ever-increasing amounts of money. What they were expected to conform to instead was working longer and longer hours in less and less fulfilling jobs in which they were sacrificing more and more of their integrity for a company that was sacrificing social responsibility in the name of profitability. In other words, being a loving person came to mean neglecting one's authenticity (including self-care), one's most cherished relationships, and societal integrity.

Most workers could not meet the requirements of excessive consumerism. The fake solution that would fill the gap between the money workers made and what they needed in order to live the 1950s version of the American Dream came in the form of credit cards. Ever-increasing amounts of consumer debt not only became acceptable but expected. This added even more stress to workers and their relationships with those they loved. The result: deteriorating health, neglected children, and skyrocketing divorce.

This is a brief portrait of how our happiness, health and prosperity, personal integrity, and social responsibility got hijacked. I give a keynote address detailing why this occurred, and also am including this discussion in my next book. The point for this article is that the Bucky Fuller quote that began this article makes it clear that he saw the handwriting on the wall. He knew that our society had developed massive integrity deficits and was doing his best to sound the alarm. Alas, too few heeded his call. Until now.

The good news is that even as close as we are today to the brink of disaster, any of us can turn all of this around in our own life. All we need to do is wake up from the spell of the 1950s delusional version of the American Dream by seeing its faulty happiness formula for what it is, and then reorganizing our lives around 3D Happiness and Integrity. The more of us who do this the sooner our society will heal the divorce that occurred between integrity, happiness, and prosperity.

What Bucky Fuller said decades ago remains as true today as it was then: "*If humanity does not opt for integrity, we are through completely. It is absolutely touch and go. Each one of us could make the difference.*" I hope you'll join me in honoring Bucky—and yourself—by helping to create the Integrity Revolution we so dearly need if we are emerge into society's next stage of evolution in a good way.

DR. DAVID GRUDER is perhaps the world's only clinical and business psychologist specializing in Integrity Intelligence and Business Integrity Management. A favorite guest of countless radio talk show hosts, *Radio and Television Interview Reports* hailed him as "America's Integrity Expert." His most recent book, *The New IQ: How Integrity Intelligence Serves You, Your Relationships and Our World,*" has been embraced across the political and faith spectrums.

The founder of Integrity Revolution, LLC, a purpose partner and faculty member with CEO Space, and a ManKind Project Elder, Dr. Gruder speaks, trains, and consults worldwide with leaders in business, government, education, and the helping professions on how to create sustainable happiness, business success, and social change without sacrificing personal integrity or social responsibility. For more information, go to www.IntegrityRevolution.com. You can read more about the hijacking of happiness, prosperity, and integrity, and simple steps you can take to reclaim your power, at the website www.TheNewIQ.com.


△ THE LAST PUBLIC STATEMENT BUCKMINSTER FULLER MADE was in answer to a question about what—after decades of accolades and achievements—he felt was most important. Without hesitating, he replied, "personal integrity."

He went on to urge that audience (and all of us) to hold on to our personal integrity no matter what and to act on behalf of all life within the context of our personal integrity. He did not tell us exactly what to do. He simply reminded us that personal integrity is key to our success as individuals and as a species.

This guidance goes hand in hand with his advice that people not believe anything that he (or anyone else) told them, but instead compare the ideas promoted by others to your personal experience. Then, you can either accept the new idea as something beneficial or reject it as the propaganda of someone seeking personal gain at your expense.

Fortunately, we now live in an era in which information is plentiful, and we can decide among information from many sources, including friends and associates. Because of our easy access to so much information, we often find that friends have firsthand knowledge of a situation that is not covered by large news sources. Friends and associates also tend to share ideas that they have thought about and researched, so their ideas often have more personal integrity.

Comparing your experience to ideas presented by others is the most viable way to decide if an idea is valid, and the thoughts of people you know personally are usually more reliable than anything that comes through mass media. This is also true of friends on the Internet after you have compared their ideas and information against your personal experience and found these friends to be honest and straightforward even when your ideologies differ. So, trusting personal experience over the ideas of others becomes a model for the integrity that is key to our individual and collective success.

It is also critical to each of us making a difference with our lives at this critical moment in time when "*it is absolutely touch and go*." When Bucky began advocating this warning in the 1950s, most people in the developed nations believed that we could go on living "the good life" forever and he was labeled a crackpot doomsayer. Today, we each have access to information with which we can prove the truth of his predictions, and many more people agree that we humans must create a sustainable culture if we are to survive and thrive.

More and more of us are realizing that it really is "*touch and go*," for our species and numerous other life forms on Earth, and we often perceive the problem as so large that we as "little individuals" can't make a difference. Fortunately, Bucky used much of his life as a "56 Year Experiment" to determine and document what one individual can accomplish that cannot be achieved by any government, corporation, religion, or other institution no matter how large or powerful. Many of the findings are recounted throughout this book.

More can be found in other material about Bucky, but the essence of his findings is that one person can make a huge difference. In fact, he found (and many of us who have followed in his footsteps agree) that the main thing that causes changes is the individual human being, especially when she joins with a few other like-minded people. As Bucky's colleague Margaret Mead said, "Never doubt that a small group of thoughtful, committed citizens can change the world. Indeed, it is the only thing that ever has."

As we stand on the brink of what Bucky described as *"Utopia or Oblivion,"* nothing is more important than individual human beings and their personal integrity. Adhering to that one aspect of human existence is key to each of us making a difference and creating the seeming elusive yet very possible *"world that works for everyone."*

# Making Our Spaceship Work

## Chapter Conclusion

 **6.7** **"We are on a spaceship; a beautiful one. It took billions of years to develop. We're not going to get another. Now, how do we make this spaceship work?"**

△ The simple question "how do we make this spaceship work?" was central to Buckminster Fuller's life and the millions of lives he touched. He continually posed it to his audiences long before ecology, peace, being green, or global abundance were fashionable or headlines were being made about global warming, the depletion of fossil fuel, and other ecological issues.

Most people did not believe him when he repeatedly reminded them that we have enough resources on our planet to take care of everybody—at a higher standard of living than anyone currently enjoys. This may still sound like illogical craziness, but Bucky was a very logical, scientific person who sought tangible proof. That is one reason he spent years discovering and documenting the simple yet profound fact that in 1976 we began doing so much more with so much less resources that we reached a point in time where we could successfully support every person on Earth.

Well before Earth Day was created in 1970 or computers could support his ecological endeavors, Bucky had inventoried all of Earth's resources and determined that humankind would soon be in serious trouble if we did not change our behavior. His initial inventory of global resources was done in the 1930s when he first discovered that any one of a diminishing number of resources including oil, water, and even clean air could trigger our eventual extinction. He also discovered that the most nonproductive use of resources is the military, and he began to champion the idea of shifting the use of resources from weaponry to livingry.

Zero attention was paid to Bucky's ideas when he first began to champion them in the 1930s. Although his calculations and predictions were accurate, most people believed that we lived on a planet of infinite resources where everything was replaceable. Still, Bucky continued on, and he did not stop at simply educating his fellow humans about this issue. Instead of giving up or slowing down, Bucky expanded his vision by creating something he called *"comprehensive, anticipatory design science."* Going well beyond basic research, Bucky employed comprehensive anticipatory design science to create solutions to problems he believed would occur fifty to one hundred years in the future.

All of his ideas had merit, but the geodesic dome became the most well known. Today, there are more geodesic domes built on Earth than any other manmade structure, and it sits waiting for the next emergency when it will need to be used. It was not, however, designed as a way to create the unique dome houses, stadiums, and the other commercial venues that dot our landscape. The geodesic dome is the most efficient structure for enclosing the most amount of space with the least amount of material. So, when we are confronted by an emergency (such as a nuclear meltdown or annual hurricane damage) that requires protecting large amounts of land or population, the geodesic dome becomes the structure used to provide a quick affordable solution.

Certainly, we have witnessed emergencies where a readily available dome would have provided a practical solution to a major problem. Had someone like Bucky been preparing for an "emergence by emergency" event such as the Chernobyl or Fujikyu nuclear disasters, they would have been ready with a huge, lightweight geodesic dome ready to contain the radioactive material. Fortunately, some visionary officials are considering using geodesic domes to do what they do best—enclose and protect massive amounts of space.

Having the vision to look into the future, Bucky made calculations to determine the feasibility of the dome. Decades ago, he proposed creating domes large enough to cover entire cities, and the city of Houston, Texas, is now considering doing just that. Such a gigantic geodesic dome would keep the temperature at a constant 72 degrees without using external fuel. It would never rain or snow inside that contained environment, and all rain falling on the dome would be captured and used. Hurricanes and other natural disasters would not cause massive destruction and fossil fuel energy use would be drastically cut by the dome's passive heating and cooling.

All too often, those of us who know of such Fuller solutions wonder why they are not being employed. The geodesic dome is only one of Bucky's ideas that can help make Spaceship Earth work for us all. The main concept he championed for most of his life was the shift of resources from weaponry to livingry (housing, education, food, transportation, and all the things that support life). Bucky continually reminded his audiences that this change was essential because we have enough resources on board our tiny spaceship to support all life if we stop trying to kill one another.

Remedies such as these have now been proven correct. We now know that shifting less than half of our global military budget to livingry will be enough to solve all humankind's physical problems. This would stop the needless starvation death of fifty thousand people every day. It would also lead to the end of global warming, and the beginning of an era in which every person has high-quality housing, food, water, medical care, etc. We would also stop destroying the natural habitat of other species so they can again flourish.

You and I would live like kings and queens of the world and so would everyone else. This is not some pie-in-the-sky dream of a crazed idealist. It is the vision of a comprehensive anticipatory design scientist who devoted most of his life to the welfare of all sentient beings. It is also a mandate that "we the people" must demand be instituted by our governments and other dominant institutions so that we can all survive and thrive on our beautiful Spaceship Earth.

## CHAPTER 7

# ACTION

## MAKING A

## DIFFERENCE

### 7.1 "Take the initiative. Go to work, and above all co-operate and don't hold back on one another or try to gain at the expense of another. Any success in such lopsidedness will be increasingly short-lived."

⚠ BUCKMINSTER FULLER GAVE THIS INSTRUCTION TO MANY of his audiences in response to the questions, "What should I do?" and "How can I make a difference with my life?" This was not, however, the philosophical meandering of someone with another good idea. This was and still is the experiential solution of a man who learned what works through trial and error experimentation during long years of being labeled a crackpot idealist.

Bucky tried to "make a living" and support his family at the expense of others, and he realized that most people play a win–lose game such as the stock market or investing in real estate with the hope that they will be winners. Such people don't, however, usually understand that—even though we label such income as earned from investments—in these "games" someone must lose a dollar for every dollar that is won. Once he was aware of the losses and suffering such initiatives caused, Bucky stopped playing games and profiting at the expense of others.

Thus, he became an advocate of complete cooperation, peace, and a *"world that works for everyone."* He also realized that the Golden Rule is more than a theoretical aphorism. It reflects the reality of cause and effect or, as some label it, the law of karma. In other words, we eventually reap what we sow, regardless if others know what we have done or not.

With that understanding Bucky resolved never to create initiatives that take advantage of another person. He always did his best to make his projects as inclusive as possible and would always tell people that he found others to be just as compassionate and caring as he was.

The issue that he found most problematic in promoting global cooperation is that the few individuals who control most of our planet's resources keep the masses in line using scare tactics and promoting constant fear. We're afraid that we won't get our share. We're afraid that "foreigners" will invade our land. We're afraid of "terrorists" and "strangers." And we're even afraid of our neighbors when we buy the scarcity story propagated by our "leaders."

Bucky's view was the exact opposite of this fear-based perspective. He did his best to never judge other people and to take initiatives that included as many people as possible for the long haul rather than the quick Band-Aid fix. He sought success for all people and would sacrifice his needs to further the welfare of another. He would also experience many losses because he chose to make long-term sense rather than make short-term money.

One such instance was the Stockade Building System. In 1926 at the age of 31, Bucky worked with the well-known architect James Monroe Hewlett to create a new modular building system that was far superior to anything on the market. Hewlett, who was also Bucky's father-in-law, and other family members and friends invested the funding for the initiative. This put Bucky in the precarious situation of being financially accountable to family and friends, and when the endeavor was another "learning experience" that failed financially, he was devastated.

It was only then that Bucky realized that he was not a good businessman. His focus was not on making money or profiting at the expense of others. As he would later recount, *"I had to decide if I would make good buildings or make money, and I chose to make good buildings."*

Stockade System buildings were affordable, easy to construct, fireproof, and well insulated. But because they could be erected by anyone, they were a threat to construction workers and their unions, especially brick masons who were a powerful force at the time. When these unions realized that Stockade System buildings would create an advantage for the average man who could use the system to build the structure for his home or business without expensive labor costs, they made sure that building inspectors caused problems every time someone attempted to erect a Stockade System building.

Still, Bucky would not allow his initiative to produce profit at the expense of other people. He wanted to build good buildings that were of use to everyone, and Stockade Systems was a financial failure until he was fired and the company was sold. From the perspective of most people, Stockade Systems was an enormous failure, and that 1927 experience was so devastating that it led Bucky to contemplate suicide.

On the verge of killing himself, Bucky had his famous epiphany when he got the message that he did not have the right to kill himself and that from that time on he would always speak the truth. Years later, he realized that the Stockade System period of his life had simply been another "learning experience." It was not a failure. It was simply another opportunity to see what worked and what did not and to focus his attention and initiatives on what worked using what he had learned to avoid another such crisis.

By using this approach to global and personal problems Bucky (and those of us who have adopted his view and followed in his footsteps) have discovered how we can each create the utopian vision of *a world that works for everyone*— including ourselves. We are also able to follow Bucky's lead that is described by Buddhists as accumulating merit. This practice of being supportive and doing good is central to many spiritual practices, and it is something needed by the world at this critical moment in our evolution.

**7.2** "There is no use talking about bright ideas. Everyone has bright ideas. There is no use talking about an artifact until we reduce it to practice, until we see whether Nature permits it, whether society permits it."

*By Dr. Cherie Clark*

GUEST COMMENTATOR

I had worked in the criminal justice system for more than thirty-six years by the time I retired in November of 2010. Throughout my career, I had applied the principles Bucky had discovered, first unconsciously, then deliberately, as I learned more, through "experience-based knowledge," about how the "omni-interaccomodative generalized principles" described in all of his talks, books, and papers truly were, "always and everywhere true."

My work kept me focused. I just kept applying the principles to develop Total Learning Environments™ based in these principles in the prison system in New York and for other states.

The two main ideas that were the basis of the model I came to name Social Synergetics™ (the science of stable relationships) provided the structure for and informed my work throughout the years I was working on application.

Bucky described the "four-cornered tetrahedron" as *"the minimum structural system in Universe."* Throughout his writings he returns again and again to the tetrahedron as the essential foundation for his inventions and "prognostications." Relative to the second key idea, in Chapter 7 of Critical Path, he says:

> *"In naval science we have four scientifically developed prognosticating arts. ... one ... that of designing and producing the generalized tools ... two ... employing the ... tools, we design ... three ... celestial navigation, ... four ... ballistics— ... interior and exterior.*

*I became gradually interested in the possibility that all variables involved in naval ballistics might be identified with all the variables operative in the most complex problems of Universe. I intuited that the combined sciences of navigation and ballistics might embrace all the variables governing Universe-event prognostication. It could be that: (1) navy yard industrialization, (2) fleet operation and individual ship design, (3) astronavigation, and (4) ballistics constituted the four 'special case' corner complexes of a generalized tetrahedral complex of variable design factors governing all human mind-controllable participation in all cosmic, alternative-intertransforming potentials."*

To understand this, I needed to draw it. It helps me to integrate ideas more easily when I do this. So I drew a simple tetrahedron and began playing around with the four corners. It looked like this when I was finished:

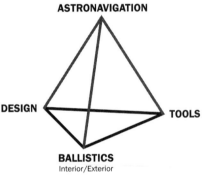

Dr. Cherie Clark

Then I tested it with every model I had studied, applied, and practiced in my private counseling practice and in the Total Learning Environments™ I designed for corrections. I realized the models studied were the tools so necessary to build the model and each also fit "the four 'special case' corner complexes of a generalized tetrahedral complex."

The second iteration became:

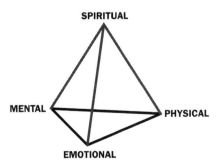

I met Lisa Matheson in 1991 and we began to work together when Lisa asked me if I had written anything about my discoveries. She kept at me until we agreed in 1993, to write about my discoveries. Together, we began writing *DOING LIFE! A Life Skills Approach to Recovery From Addictions*, with contributions from a colleague, Mary Bogan. The more I saw the exact connection between the six motion freedoms of the generalized principles and the Twelve Steps to recovery, the more clear I became that the two independent discoveries exactly paralleled each other.

The longer I worked with what was the beginning of the Social Synergetics™model, the more I could see that the 12 Steps were "special case truths" of the Twelve Degrees of Freedom Bucky spoke about as "*omni-interaccomodative generalized principles.*"

After completing and testing *DOING LIFE!* Lisa insisted that I keep writing and exploring this idea. We wrote *S.M.A.R.T. CHOICES!* Also based in all of the models I had studied and then it was time to formalize all of it.

I started writing my doctoral dissertation, *Twelve Degrees of Freedom: Synergetics and the 12 Steps to Recovery*, on nights and weekends from 1996–2001, while still working full-time to apply the model to the TLE™ in prisons in New York and around the country. I had lots of current examples of how Synergetics had supported the development of the Social Synergetics™ model.

There were countless times that I'd think, "I know Bucky said ... about this." I could hear him saying it, and wanted to make sure I had the exact quote. I'd reach over to my pile of "Bucky books," pick up one and open it to the exact quote I was looking for. At first, it startled me. Then, it became a game. Usually around 3 a.m. I would hear Bucky's distinctive voice, with a suggestion for how to more clearly respond to a question I was wrestling with earlier. He was right beside me the whole time I was writing, more than thirteen to eighteen years after he had departed this life, guiding, coaxing, teasing me to understand what he had said.

Now more than ever, I believe Bucky's time has come. He once said he worked so long into the future so that people would not misinterpret his words. I believe his extensive body of work is even more viable today, more necessary than ever. Bucky wanted everyone to study Synergetics, convinced that his discoveries would help us to solve the "most complex problems of Universe." Let us continue to do just that.

DR. CHERIE CLARK retired from the New York State Department of Correctional Services in November 2010 after forty-three years of state service, thirty-six in corrections. At the time of her retirement, she was Director of the Shock Incarceration Program and Willard Drug Treatment Campus. The Network Program was the first variation of this intensive educational and life skills intervention.

Due to the unprecedented success of Network, she was asked to design and direct Shock Incarceration, a six-month intensive treatment program for non-violent felony offenders. At the time of her retirement, there were more than forty thousand graduates and documented savings of more than $1.4 billion. Dr. Clark is an author, philosopher, and futurist, dedicated to having everyone understand the elegant simplicity of Synergetics as a foundation for solving all problems. See www.socialsynergetics.com for more information.

BUCKMINSTER FULLER WAS EXTREMELY RESOLUTE ABOUT the fact that action is critical to success. Again and again, he would remind audiences that he was "just an average healthy man" and that they could achieve much more than he had done if they took the initiative. Such initiatives were a key element in his being able to accomplish so much and make such a differences in one lifetime.

He continually reminded people that information is constantly flowing throughout Universe, and sometime a person will "tune in" to that information and have what they believe is a bright idea. These flashes of insight are what Bucky labeled "cosmic fish," and he was constantly "cosmic fishing" by recording the fish/ideas before he forgot them.

Once they were recorded, he began the process of evaluating them to find the ones that would make the most difference for the greatest number of people. Then he began the process of reducing the best ideas to practice, usually by constructing a physical artifact—be it a new form of geodesic dome, a three-wheeled car, or a book. This was the test phase mode.

Bucky knew that nothing could exist unless Nature permits it. He often said that there is nothing artificial in the world. Either Nature allows it to exist or it does not exist. And he knew that if his cosmic fish artifact was something that was beneficial to Universe and Nature, he would be supported in the creation of the model of the artifact and the artifact itself.

Time and time again, he proved this strategy to be valid and viable. The Dymaxion Vehicle is a perfect example of reducing an idea to a genuine physical entity. When he first came up with the idea in 1931, it was so outlandish that he was considered a crackpot visionary, but Bucky persisted and built a scale model to show people. He also began speaking about both his futuristic Dymaxion House and Dymaxion Vehicle to anyone who would listen.

Eventually, people began to marvel at the models and renderings of both the House and the Vehicle. Then, a few days after President Roosevelt temporarily ordered all the banks shut down, a man who had heard Bucky speak about his idea and models gave him a suitcase full of cash to build a prototype.

One significant element of this manifestation that many people overlook is that it did not just happen overnight or the moment Bucky had the idea. The major support he needed did not even show up during the years he devoted to making presentations about the practicality of his ideas and models. Bucky had to persist. He often experienced low points and times when he felt that his ideas would never come to fruition. Still, he maintained some semblance of faith, knowing that

Universe and Nature always support what needs to happen and exist.

Decades later, he encouraged his protégés, students, and audiences to do more than just talk about their good ideas. He urged them to produce some form of them and share that model, book, or other artifact with others. Only when engaged in that process can we know whether Nature and society will permit it. And only then can we truly know and appreciate the natural order of all things, including humans and the "cosmic fish" bright ideas that constantly swim through our minds.

**7.3** "What can a little man effect toward such realizations in the face of the formidable power of great corporations, great states, and all their know-how, guns, monies, armies, tools and information?

The individual can take initiatives without anybody's permission. Only individuals can think and can look for the principles manifest in their experiences that others may be overlooking because they are too preoccupied with how to please some boss or with how to earn money, how to take care of today's bills. Only the individual disregards his fears and commits himself exclusively to reforming the human environment by developing tools that deal more effectively and economically with evolutionary challenges.

Humans can participate—consciously and competently—in fundamental ways, to changes that are more favorable to human life. It became evident that the individual was the only one that could deliberately find the time to think in a cosmically adequate manner."

By *Peter Meisen*

GUEST COMMENTATOR

The following is a chronology of turning an idea into action. It outlines how we created the Global Energy Network Institute (GENI).

*Early Inspiration*

I remember very clearly the first time that I heard Bucky speak. It was in 1971, and I was invited by my Dad to hear Bucky deliver one of his infamous four-hour "Tours of the Universe" in Oceanside, California. I was enthralled, dumbfounded, but couldn't actually tell you what he had said. As a high school senior at the time, I knew what Bucky was saying was important but little did I know how this experience would change my life.

The next time I met Bucky wouldn't be for more than ten years when I was invited to the Hunger Project Board meeting where Bucky was a guest speaker. Older now, I was able to better understand his global vision and extended answers. It was at this meeting that someone asked him, "Bucky, what should we do?" and in typical fashion, he encouraged us to look for what was wanted and needed in the world, and he suggested we read his new book, *Critical Path*.

It wasn't until three years later while on vacation, when I was looking for the next project in my life, that I finally got around to reading the book. On page 206, Bucky states that the *"global energy grid is the World Game's highest priority objective."* I realized that this strategy was most wanted and needed, and I decided to take a stand.

*Initial Stages*

Needless to say, the scope of the project scared me to the bone. With my engineering background, I had a core understanding of the importance of electrical energy to sustain

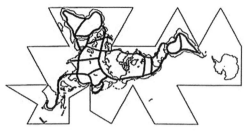

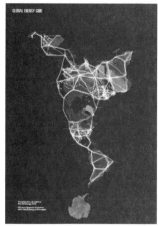

Fuller's Dymaxion Power Grid displayed on his Dymaxion Map.
The Global Energy Network Institute (www.geni.org)

modern society. In *Critical Path*, Bucky explained how the linking of abundant renewable energy resources between nations and continents would reduce pollution and population growth while increasing the quality of life for all.

I began to re-educate myself, scouring libraries (there was no Internet), and digging into the Buckminster Fuller Institute archives in Los Angeles. I had "played" Bucky's World Game™ beforehand and used its mission statement as my creed:

*"How do you make the world work for 100% of humanity in the shortest possible time through spontaneous cooperation without ecological damage or disadvantage to anyone?"*

My first presentation was to friends around the kitchen table with a Dymaxion Map and pen drawing of the grid concept. There was no budget, just stationery, business cards, and a small mailing list.

The original three-year business plan was to educate leaders and utilities about this solution, then generate agreements between nations, with the ultimate end goal of actually building it. How naïve I was! Energy systems and electric grids are the most complex machines ever created, and naturally, tend to change very slowly.

## Official Formation

GENI was founded as a nonprofit in 1991. Its first office was a one-room, donated space in San Diego, where we worked on an IBM 286 laptop. That same year in Manitoba, Canada, we organized an international workshop on "The Limits of Long-Distance, High-Voltage Power Transmission and the Corresponding Economic, Environmental, and Social Political Implications." Thirty-six experts from around the world attended and gave their credential to the economic and social benefits of interconnecting grids between regions and nations. We created a video, "A Win-Win Solution," which led to endorsements and allowed us to develop our first website in 1995.

I recall the first conversations with executives and engineers who would listen politely, nod, and respond with "good luck." They knew the challenges and hurdles much better than I and at that time the Berlin Wall meant that East and West would never connect.

## Moving Forward

Buckminster Fuller was a visionary engineer ahead of his time. Today, when we speak about "linking renewables," the prevailing response is "of course, we need to do that." The benefits to jobs, domestic energy, and climate change are clear. The problem is that we are moving at a half step when we need to be moving double time. Bucky gave us a road map in *Critical Path* and *Operating Manual for Spaceship Earth*, now it's our job to get it built. Bucky often reminded us,

> *"Now we've reached the point of discovering that muscle is nothing, mind is everything. Evolution is integrating us and we're no longer so remote from each other. Clearly we are here to use our minds, to be information gatherers in the local universe, problem solvers in relation to the*

*maintenance of the integrity of the eternally regenerative universe.*

*Muscle is nothing; mind is everything. But muscle is still in control of human affairs. In about ten years, if we come out with muscle in control, we will have chosen oblivion; if we come out with mind in control, it's going to be utopia and eternity. Yes, we do have the option to make it, but it's absolutely touch and go, a matter of the integrity of every human being from now on."*

PETER MEISEN founded GENI in 1989 to conduct research on and educate world leaders about the strategy of linking renewable energy resources around the world. He helped to create the KLD/MSCI Global Climate Index to foster investment in climate solutions. Meisen led the development of the World Resources Simulation Center to visualize and accelerate sustainable solutions. He is an internationally recognized speaker on the global issues of renewable energy, transmission, and distribution of electricity. His work can be found at can be found at www.geni.org and www.wrsc.org

△ BUCKMINSTER FULLER CONTINUALLY REFERRED TO HIMSELF as "the little man," and he would frequently tell his audience that the most important thing about him was that he was an "average man." This was not the put-down of an insecure man, but rather the suggestion that anyone could achieve what he had accomplished and more. It was also the approach he used to introduce people to the things he had discovered that would make the biggest impact with the most people in the shortest period of time.

When Bucky was initially formulating what would eventually become his "56 Year Experiment" to determine and document what one individual could achieve, he had no resources, had created no positive results, and had no formal education beyond high school. In other words, he was a failure in the eyes of society, and he almost fell into the trap of agreeing with that perspective. Fortunately, he had a young daughter and wife to support, and he looked to see what he, as an average little man could do that could not be done by any organization, institution, or corporation no matter how large or powerful.

It was then that he realized that no organization or corporation had ever thought or taken action. Only individual humans can think, discover Nature's generalized principles and incorporate those principles into action that serves all people. This simple strategy became his trimtab to make the most difference with the least amount of effort.

Regardless of the propaganda espoused by large corporations, governments, religions, and other institutions, we can all follow this simple path in creating the seemingly elusive "world that works for everyone" right here and right now. Only we individuals can think and take action. No corporation will ever generate a single thought—much less invent the next great app, computer program, or ground transportation vehicle. No religion will ever come up with a single inspirational aphorism. And no government will ever shut down a single military facility. These things are all initiated and accomplished by individuals.

Sometimes these individuals work together to accomplish a task, but the initial idea and impetus always and only arises in the mind of a single person. It may appear as a thought to several people at the same time, but rarely does more than one person take the needed action to manifest the idea into physical reality. Bucky often shared his unique perspective on why this happens when he said:

*"All the knowledge in Universe may have been known to various people at other times. However distant or remote any information signal is, it has to just go on forever unless it is interrupted."*

The interruption of that signal happens when it appears as a thought or what Bucky labeled a "cosmic fish." People like Bucky who are practiced at "cosmic fishing" record these cosmic fish the instant that they appear. He always carried a small notebook for such moments, and many of them resulted in his most famous inventions and books.

This phenomenon and the ability to make enormous changes in our world are available to everyone. It can't be taken from you, but you must cultivate it if you want to be a fully functioning and participating crewmember on board Spaceship Earth.

You have to look beyond making the next dollar, driving a newer car, or building a bigger house and develop a perspective of universal competent curiosity. You have to realize that only individuals can take action, and that if you do this from the perspective of servicing the whole of humankind, you will be supported and rewarded. That's the simple law of cause and effect, and it can lead us all to lives of contribution, satisfaction, and freedom well beyond what most of us can imagine.

**7.4** "My working assumption is that we are here as local Universe information gatherers. We are given access to the divine design principles so that from them we can invent objectively the instruments and tools that qualify us as local Universe problem solvers in support of the integrity of an eternally regenerative Universe."

*By Michelle Levey*

GUEST COMMENTATOR

I had the great good fortune to personally meet Buckminster Fuller at the first World Symposium on Humanity in Vancouver, B.C., in the fall of 1976, and was so moved, inspired, and expanded by his presence and by what he had to share that my life was forever changed by that meeting. This statement has always been one of my favorite Bucky quotes and one that I first encountered at that time.

Posed as a "working assumption," Bucky invites us here to shift our view to behold ourselves as embodiments of the Universe, discovering and awakening more fully to itself through our being. The portal of our creative intelligence is open to receiving "divine design principles" to fulfill our role as *"local Universe information gatherers"*—if we are able to develop the right blend of clear presence, depth of reasoning, intuition, and motivation that will enable us to realize this innate potential and "qualify" as innovative problem solvers in service to the Whole.

What is your assumption about why you are here and why humans live in this world?

As we refine our capacity to live and serve in support of the integrity of an eternally regenerative Universe, we awaken ever more deeply to the experience that the upwelling flow of life experience that breaks upon the shore of our being moment-to-moment, breath-to-breath, heartbeat-to-heartbeat, and mind moment-to-mind moment is the upwelling revelation of life presented uniquely to and through each of us in every precious moment. As we learn to align and attune ourselves ever more deeply to be present to what is expressing itself as universal design principles at play, we become ever more capable of discerning the true nature of reality and living in creative, inventive ways that are congruent and in harmony with it rather than opposed to it.

Can you imagine how different our world would be if we were to develop our capacity to truly understand these divine design principles that determine the nature of life, the Universe, and ourselves, and to live in support of its integrity and wholeness? Can you imagine how different our world would be if we learned to free ourselves from what Einstein called the "optical delusion of consciousness" that leads us to view ourselves as separate from all creation, and to instead, "widen the circle of our compassion to embrace the whole of nature in all of its beauty?"

As Bucky reminds us—we have access to these *"divine design principles."* They are not hidden from us. They live within each of us and throughout the Universe.

Will we seek out the wisdom teachings and practices necessary to discover and fully fathom those principles, and then align ourselves with them so that we honor and support them in how we live our lives and design our workplaces and communities? Or will our species continue to compromise them out of ignorance, and thereby act in ways that create immense suffering for ourselves, others, and generations to come?

A good practice is to check in with yourself frequently throughout the day, asking:

- Is my mode of being in this moment in support of the "eternally regenerative Universe"—or not?
- Are the divine design principles of the Universe clear to me in this moment?
- Are the choices I am making and the actions I am engaged with in harmony with and in support of the whole field of complex relationships in which I am embedded in this moment—or not?

In this wonderful quote, Buckminster Fuller gives us the gift of a great spirit of optimism and a tremendous affirmation of our co-creative potential. It is up to us to choose to accept this remarkable invitation, to develop our wholeness, deep listening, and problem-solving capacity, and to live in ways that are in harmony, alignment, and support of it—in service and for the resilient benefit of all.

MICHELLE LEVEY is co-founder of InnerWork Technologies Inc.; WisdomAtWork. com; and the International Institute for Mindfulness, Meditation, and MindBody Medicine. Michelle also serves on the Advisory Board of the International Institute for Compassionate Cities and stewards the Kohala Sanctuary, a permaculture-designed learning center on the Big Island of Hawaii. She is the co-author of many books and audio programs including *Living in Balance*; *Luminous Mind*; *Wisdom at Work*; and *The Fine Arts of Relaxation, Concentration, and Meditation,* and blogs for the *Huffington Post* in their Healthy Living—Spirit section. Michelle's work with leading organizations around Spaceship Earth draws insights from science, spirituality, and medicine, and offers an integrative approach to change resilience focusing on personal, community, and organizational sustainability and thrivability. For more information about Michelle and her work go to http://www.wisdomatwork.com.

⚠ BUCKMINSTER FULLER DEVOTED A GREAT DEAL OF TIME TO contemplating why he (and all humans) was here on Earth. After serious consideration, he concluded that humans are here to gather information and solve problems locally in ways that are sustainable. In other words, we humans discover new information in order to create sustainable solutions to problems in this "local" region we call Earth.

Working from that assertion, it's fairly easy for a person to decide what to do in any given moment. All she needs to do is to stop before taking any action. Then, she looks to see if her proposed action will gather information and solve problems in a way that supports sustainability. Fortunately, we each have unlimited access to what Bucky called the Divine Universal Mind, and that mind can provide us with a quick answer to the issue.

For most people, that answer emerges as an intuitive idea or thought. It seems to simply arise from nowhere, but that is generally not the case. People can, in fact, develop their capacity for intuition. Bucky himself felt that intuition was so important that he named his beloved sailing sloop *Intuition*. He also almost always followed his intuition even when the suggested actions seemed illogical.

If we really consider what we do on a daily basis, we find that our most productive and rewarding actions are usually the ones that fit into the above definition. In other words, we gather information and solve problems in a sustainable manner. Those activities cannot help but contribute to other people, all forms of life, and the welfare of our planet.

The other very important element in this quote and our ability as individuals to make a difference is a single word that most people gloss over. It's not divine or design principles or Universe or integrity. The pivotal word that is extremely important at this juncture in human evolution is "qualify."

The question Bucky asked himself and his audiences throughout his "56 Year Experiment" is *are humans a worthwhile to Universe experiment?*" In other words, do our actions create results that qualify us for our universal mission?

When asked if he thought we qualify and would survive as a species, Bucky usually answered, *"it's touch and go."* Only in the final weeks of his life did he declare that he thought humans would in fact make it. And that statement was followed by a warning that as responsible individuals we should not let up and hold on to our personal integrity no matter what.

### 7.5 "You never change things by fighting the existing reality. To change something, build a new model that makes the existing model obsolete."

*By Bill Kauth*

GUEST COMMENTATOR

To me, this is the classic Bucky quote. It's akin to Margaret Mead's "Never doubt that a small group of committed people…" or like the fellow who said "Where there is no vision, the people perish"—there are some wisdom quotes that are simply true over time. This quote graced the inside cover of my new book as it has so many other books. It is an all-time great quote.

What makes the truth of Bucky's observation so poignant and has kept me dumbfounded for decades, is that most of the world's people, including myself, seem constantly caught in fighting the "existing reality." Why do we do this obviously counter-productive behavior?

After years of wondering why we keep fighting, an answer has emerged. Brain science and research now have shown that we humans have what we might call a "negative bias." Our brains are wired to look 95% of the time for danger. For hundreds of thousands of years the ancestors who were not constantly vigilant failed to propagate that species.

So here we are in the most abundant time in all of human history, constantly ready to fight for survival. Somehow Bucky's inner wiring was designed to look for the bright possibility. For him the fun of it seemed to be the creative act of building the new model. I like to imagine Bucky was simply allowing himself to be led by his joy. He was doing what Joseph Campbell suggested as "follow your bliss."

The soulful implications of the joy/bliss invocation take us down to meet the divine. I like to imagine that Bucky knew that the building of a new model of whatever was that moment when the hand of God becomes your hand. If god is universal or omnipresent we can see the divine plan unfolding everywhere. And if I am in alignment with sacred evolution of God expanding, becoming more complex, I always feel a feeling we might call "joy." This joy is an inner knowing of the rightness of all things. It just feels joyous. We might imagine this feeling as God's reward to us for doing the right thing, or helping her grow and create beauty.

Now, I doubt that Bucky has ever been accused of being a saint, but there was something decidedly transcendent going on. As a prophetic character he was well ahead of the dominant worldview. Bucky knew the GRUNCH and how that self-serving worldview perpetuated the fight, always "us against them." He must also have seen how that system crushes true innovation. He is calling for us to leap over the snake pit of endless conflict and use our great creative potency to build the new world allowing the old order to give way.

Now, the collapse of old systems seems stunningly obvious to most everyone. This "in your face" awareness makes Bucky's admonition even more poignant as we can see that the "obsolete" system coming apart, just begging us to step up and build the new model. And as Paul Hawken has demonstrated in his book *Blessed Unrest*, people all over the planet, by the millions, are stepping up. New models for water, food, economic, social, and energy systems seem to be bubbling out of the fertile field of human spirit and imagining.

BILL KAUTH is a visionary and social inventor. In 1984 while working as a psychotherapist and business consultant, Bill conceived and co-founded the "New Warrior Training Adventure." Now with fifty thousand graduates, it's presented by The ManKind Project with centers in forty-two regions in eight countries around the world. In 1992 St. Martin's Press published his book *A Circle of Men: The Original Manual for Men's Support Groups*. Over the last two decades Bill also co-founded the Spiritual Warrior, the Inner King Training, and in 1995 the Warrior-Monk Training Retreat. In his service to the ManKind Project as "Visionary at Large" Bill intensely studies the state of the world. One clear conclusion: everyone wants community yet no one knows how to build it. Heeding the call Bill has, with his wife Zoe Alowan, published *We Need Each Other: Building Gift Community* and delivers seminars in several countries.

*By Rick Ingrasci, M.D., M.P.H.*

GUEST COMMENTATOR

From 1974 to 1980 I had the honor and privilege of hosting Bucky at a series of public seminars in Boston sponsored by Interface, a nonprofit holistic education center. We called these seminars "Being with Bucky" because while Fuller's contributions can most readily be recognized through the artifacts he created—the geodesic dome, synergetic geometry, Dymaxion Maps, the World Game, etc.—the deeper essence of Bucky's contribution was the quality of being from which his inventions and discoveries came.

At the heart of Fuller's "beingness" was a loving man with a passionate belief in the goodness of people and our capacity to create a world that works for all. Bucky saw the Universe (he always spelled it with a capital U) not as a collection of things but a harmonious, all-embracing set of relationships, eschewing our tendency to see the whole in terms of parts instead of parts in terms of the whole. Fuller, like Einstein, was a scientist with a whole-systems, essentially mystical worldview that is reflected in this statement:

> *"When you see a fresh stream of water working your way, if you scrape the earth a little, the water will run in your preferred path. Humans can participate consciously and competently in fundamental change in ways that are favorable to all life. I could see that where the circumstances are favorable, children can grow while continually regenerating love, affections, thoughtfulness, and competence. I was thus inspired to committing myself to discovering generalized, cosmic principles and reducing them to special case physical tools and practices."*

As a psychiatrist interested in personal and cultural transformation, I was particularly intrigued by Fuller's insights into creating change as seen in this statement:

*"You never change things by fighting the existing reality. To change something, build a new model that makes the existing model obsolete."*

In addition to his own inventions, he would often use simple illustrations of "doing more with less" (ephemeralization), synergetics (the whole is greater than sum of its parts), and design science. For example, if people are starving on one side of a raging river, unable to cross to reach ample food on the other side, then don't argue about the best swimming strategies, build a bridge. No discussion, no fuss—people will use it because it's a better solution to the problem.

I reflected on Bucky's *"Build a new model"* insight for many years as I attempted to answer the question "How do we facilitate the transformation of human consciousness from a materialist/reductionist worldview to a holistic/integral worldview?" The critical need to do so is clear: Everyone has a story about the nature of reality, or a worldview, that shapes our capacity to understand the world. Our worldview is therefore a core underlying force determining our attitudes, values, beliefs, and behaviors.

As Albert Einstein said, "No problem can be solved from the same level of consciousness that created it." To change the world, we must collectively change our minds. The critical need is to shift from an egocentric way of being (with its incumbent fear, alienation, loneliness, and violence) to an ecocentric consciousness rooted in a deep sense of the interconnectedness of all life. The great need is for humans to recognize and radically increase their connections to each other and the whole Universe.

My answer to Bucky's challenge to *"Build a new model"* came to me in the form of a slightly tongue-in-cheek metaphor: "If you want to create a new culture, throw a better party!" It seemed to me that our current way of being together is rooted in joyless competition and mechanistic separation, which is not our true nature. What's needed is a more convivial, cooperative, creative, and compassionate way to organize ourselves on planet Earth.

The metaphor of a "Better Party" speaks to this new model because it is grounded in the basic design of human nature. The latest neuroscience confirms what is intuitively obvious, that humans are social animals whose very survival depends on living in cooperative communities with deep social bonds and healthy emotional attachments. We also know that creativity and play give humans a powerful evolutionary capacity for adaptation and change, and that true happiness can only be experienced when we serve something greater than ourselves.

Jeremy Rifkin has summarized this perspective in his latest book, *The Empathic Civilization,* where he writes, "A new rendering of human history and the meaning of human existence ... focused not on the conflicts, antagonisms, and power struggles that have marked human progress, but on the empathic evolution of the human race and the profound ways it has shaped our development and will likely decide our fate as a species."

The Better Party model for cultural transformation has been the focus of my work for the last thirty-five years. I've studied what Barbara Ehrenreich has aptly named the "History of Collective Joy" and discovered that the human impulse for communal celebration is built into the fabric of our being.

To this end I've been exploring the power of creativity and play at an annual gathering of innovative leaders called the Hollyhock Summer Gathering that I've been convening since 1986 (see www.hollyhock.ca). One hundred twenty social artists and entrepreneurs from diverse cultural and intergenerational backgrounds come together for one week at the Hollyhock Retreat Center on remote Cortes Island in British Columbia to experience a community of practice rooted in love, trust, play, compassion, and authenticity. It is, indeed, a Better Party ... in service of the whole.

A group of colleagues and I are currently launching the Whidbey Geodome Project, a social experiment in transformational education using a portable inflatable Geodome as a central feature for immersive multimedia storytelling (see www.newstories.org). Our goal is to immerse participants in the Universe story, the Earth story, the Human story, and local bioregional stories that show the interconnection and beauty of all life. We combine these peak experiences in the Geodome with an experiential learning process outside the dome designed to activate people's ability to engage in building sustainable communities. We are developing specific curricula (rooted in Fuller's design science) for education, business, and public settings.

Clearly our capacity to create social capital—i.e. relationships based on trust and reciprocity, rooted in love, empathy and compassion—is the deeper truth about who we are as a species, and also the most adaptive response to the evolutionary crises we face on planet Earth. Why not try "Better Parties" as a new model? It's serious fun!

RICK INGRASCI has a rich background in psychiatry, holistic medicine, community development, and social entrepreneuring. He currently practices life coaching, mainly with leaders of non-profit organizations. Rick co-founded Interface (Boston's largest holistic education center), the American Holistic Medical Association, Physicians for Social Responsibility, and Hollyhock Retreat Center, where he has been convening an annual Hollyhock Summer Gathering since 1986 (see www.hollyhock.ca). He co-authored the bestselling *Chop Wood, Carry Water: A Guide to Finding Spiritual Fulfillment in Everyday Life*. He is the Director of the Whidbey Geodome Project, a nonprofit transformational education program that utilizes an inflatable immersive multimedia Geodome to aid in the development of a more ecocentric worldview (see www.newstories.org).

*By Michaela Miller*

GUEST COMMENTATOR

Watching Tropical Storm Fay crisscross Florida four times on a battery powered TV and through our windows was scary. When our bulkhead finally breached on August 22, 2008 and flooded our home—I knew our reality was destroyed. We salvaged four pieces of furniture and one pair of shoes each.

For six long months we fought with FEMA and the insurance company and finally settled. We had an exquisite riverfront property with a beautiful view and decided to rebuild. We deconstructed the old house and rebuilt on nine-foot steel stilts (Category 5 proof) utilizing every energy efficient, sustainable, and environmentally responsible product we could find and afford. We accomplished even more than we had originally thought possible!

We diverted 303 tons of debris from the landfill by recycling, reusing and repurposing construction materials. Our new home is the first in northeast Florida to achieve Platinum LEED (Leadership in Energy and Environmental Design), Gold Water Star and Energy Star status. We chose to break the existing model of home building and now speak regularly to civic, environmental, and building professionals about the need to protect our resources and planet.

MICHAELA MILLER founded Video Law Services, Inc. in 1985, after ten years as an award-winning television journalist. Her experience at television stations in Houston, Boston, Providence, and Jacksonville includes personal and team responsibilities as a reporter, anchor and producer.

Michaela and her husband, Steve Sadler, became unwitting leaders in the Green Building during Tropical Storm. The couple built the first Platinum FGBC (Florida Green Building Coalition) home in NE Florida. With a HERS rating of 18, it is one of the most energy efficient homes in the U.S.

⚠ BUCKMINSTER FULLER WAS CONSTANTLY BUILDING NEW "models" of almost everything. This was his strength as well as something that caused him to be less well known than he could have been because people could not define him. He was a comprehensivist, interested in everything and working on many different projects. Thus, people categorized him as an architect, philosopher, poet, designer, writer, inventor, sailor, speaker, environmentalist, etc., etc, and his books are found in several different areas of any library or bookstore.

In each of these areas, Bucky created models or artifacts to demonstrate the obsolescence of the old way and provide a new solution that people could observe. Even his poetry was unique, sometime filling an entire book with one epic poem that people studied rather than read.

Other important models include his books and oral teachings. These offer the comprehensive solution of shifting from a competition-based weaponry society to a cooperation-based livingry culture. That is the model for creating the global society labeled *"a world that works for everyone."*

People are often overwhelmed when considering this model-based method of creating change. They think that a single individual can't create a model that will result in societal shifts but, as Bucky demonstrated (using his life as the model), nothing could be further from the truth.

Every change begins with one person's idea followed by her taking action to build a new model. This occurs because two or more people cannot come up with the exact same idea at the same moment in time. People can collaborate on an idea once it has been established in the form of a "model," but change always and only begins with one individual. If that individual's idea and model are viable, needed and exciting, then others join in to make the existing model obsolete.

When a person realizes that she can, in fact, effect change, she needs to create a model will make the current reality obsolete. In his models Bucky almost always looked to his immediate environment and Nature. Within his manmade environment Bucky found plenty of things that needed to change as well as Nature's examples of the best way to solve those problems.

There are plenty of things that need to change, both large-scale and smaller. It might be a relationship or it might be the recycling of all our waste. Whatever the case, consider doing as Bucky did and simply begin.

You will have learning experiences (often labeled mistakes), but those are simply opportunities to course correct while moving forward. If you have a viable, needed model, it will attract what is needed to it. All you have to do is steward that idea through to completion or to the next phase, which may be taken over by another person or group.

## 7.6 "The best way to predict the future is to invent it."

*By Hunter Lovins*

GUEST COMMENTATOR

I recently used this quote in a presentation to representatives of a conservative city uncomfortable being able to access federal money because they had labeled their proposals as "sustainable development." One of the developers leveled on me, observing that I was either exceedingly pretentious calling him Bucky, or I knew Mr. Fuller.

Grinning, I answered that it could be both, but yes, I knew Bucky. And I treasured greatly the times I enjoyed with him.

In many ways the father of the discipline we now call sustainability, R. Buckminster Fuller envisioned a world that works for 100% of humanity. We now know that achieving this goal is not only possible and preferable—it is the only way to avoid catastrophe.

In our recent book, *Climate Capitalism*, my co-author Dr. Boyd Cohen and I describe how entrepreneurs, communities, and companies are prospering by using resources more efficiently, by wholly redesigning how they make and deliver products and services, and by managing their institutions to be restorative of human and natural capital, the forms of capital that Bucky recognized as underpinning all life and thus all economic systems. Bucky understood the urgency of transforming our economy to do this.

I remember sitting with him in the Denver airport waiting for a plane that would bear us to one of John Denver's periodic gatherings of change agents. Bucky spread his Dymaxion map on our knees, pointing to the Himalaya, and remarking (in 1981) that climate change would melt the glaciers. *"They water 40% of the people on Earth,"* he observed, making the prescient point that we simply must avert climate change.

We are inventing the future. It's just not the one that any of us want. It's time, again, to return to Bucky's wisdom in *Operating Manual for Spaceship Earth.*

Bucky was fond of saying, *"If the success or failure of this planet and of human beings depended on how I am and what I do ... How would I be? What would I do?"*

In our book, Boyd and I take a crack at answering that, describing how to reinvent capitalism now, so many years after Bucky suggested that it might be a good idea. We suggest inventing an economy that runs not on what Randy Hayes calls "cheater capitalism," but on full cost accounting, on fully valuing the contribution of human and natural capital to our economy and well-being, and, as Bucky once said, on love which he defined as *"... omni-inclusive, progressively exquisite, understanding and tender and compassionately attuned to the other than self."*

Not for the last time, I wish that the world had listened more closely to Bucky when he was with us. *Let's do it now.*

HUNTER LOVINS is president of Natural Capitalism Solutions (natcapsolutions.org), professor of sustainable business management at Bainbridge Graduate Institute (BGI. edu), and chief insurgent of the Madrone Project (madroneproject.com).

△ WE OFTEN HEAR SELF-APPOINTED PERSONAL GROWTH GURUS telling us that we create our own reality, but none suggest that we can predict our own future by inventing it as Buckminster Fuller both advised and demonstrated. The future he sought to invent was his legendary *"world that works for everyone,"* and he was quite successful at helping people to appreciate the possibility of this being a viable concept.

Many of us are now beginning to realize that a *"world that works for everyone"* is not only feasible, but that it must come to pass if we are to survive and thrive. We can no longer continue acting in the unsustainable manner that got us into the mess that we find our planet and most life on it, and we have to invent a future that includes the welfare of all sentient beings if we humans are to continue, grow and evolve as individuals and as a species.

Win/win scenarios are a necessity. War is obsolete, and peace must become normal and natural. Politics and politicians are obsolete. Cooperation must become the norm and replace competition. Love must replace fear. And the feminine must become balanced with the currently dominant masculine.

These are just some of the primary elements we must rediscover in the near future if we are to choose utopia over oblivion, and, as Bucky warned decades ago, the choice is just that drastic. We're a species on the edge of extinction, and we'll most likely destroy many other forms of life in the process of our own destruction if we don't wake up and change soon.

This, however, does not have to be the case. Many people, including those who have generously written essays for this book—are working to create positive solutions to some of our most severe challenges. They are the ones who predict that we can invent a positive future by creating what "we the people" want rather than something being promoted by those claiming to be our "leaders." Most of the "1% leaders" are actually selfishly accumulating resources for themselves and their small circle of family and friends.

The future that the "1% leaders" want to invent does not include the whole or a holistic perspective. They continue to believe that we live in a "you *or* me" world when we have actually transformed our physical reality so that we can now live in a "you *and* me" world. That's the future that Bucky envisioned, and it's the future I want to live in—now. Please join those of us who have stepped across to this new vision where we continue to invent an endless potential of abundance, peace, love and joy.

# TAKE ACTION AND DO SOMETHING!!

## CHAPTER CONCLUSION

 **7.7** **"We are not going to be able to operate our Spaceship Earth successfully nor for much longer unless we see it as a whole spaceship and our fate as common. It has to be everybody or nobody."**

⟁ I END THIS CHAPTER AND *A FULLER VIEW* WITH ONE MORE reminder of Buckminster Fuller's challenge to us all. He often said that the first truly global holiday should be celebrated on July 16 because on that day in 1969 modern man initially landed on the Moon. July 16, 1969, was the first time in recorded history that humans could look out from another natural site in space and see our fragile little planet floating in the blackness of space.

Just as important was the fact that that panoramic vision was not limited to the two astronauts on the Moon. Modern technology allowed billions of us to watch the Moon landing and Moon walk from the comfort of our living rooms.

All we had to do was tune in with our fellow travelers on Spaceship Earth, and we, too, could view ourselves on a planet devoid of the boundaries that we were all taught divided nations, cities, and individuals. With that single sight, we could begin to appreciate the fact that we humans are one people who have a common fate, and we began to have a visceral feeling for the inevitability of a paradigm shift from a "you *or* me" world to a "you *and* me" world.

Coming from that context, success is not limited to a small segment of the population. *"It has to be everybody or nobody."* Either we all survive and thrive or nobody does. Your income or physical assets are not important if there is not clean water to drink or fresh air to breathe. Your hoarded stockpile of food and water is of little use if humans can no longer find fresh food and water on our Mother Earth. And your governments cannot protect you from "foreigners" if a few still-living humans turn into marauding crowds in search of scarce resources.

Knowledge of such potential disasters helps to eliminate them because once we are aware of something, we can consciously act to change it. None of the above scenarios needs to happen. When we recognize that we are living on an abundant planet that has more than enough resources to support everyone, we free ourselves from fear and greed. All we need to do is shift our perspective from false scarcity to the genuine abundance and from weaponry to livingry.

Survival and flourishing then becomes the norm for all people and sentient beings. Still, even the most magnanimous spiritual being cannot survive without this shift. You might consider yourself to be a spiritual loving person, but you can't simply sit on your cushion meditating and expect things to change. You, like all of us, have to take action. You have to find the things near you that need to be done and do them on behalf of yourself and others. You have to share the message that we humans are a species on the verge of extinction with a sense of possibility because without uncontaminated food and a stable climate, no humans will survive.

One critical aspect of this change to always keep in mind is that it will take everyone's input, gifts, and cooperation for our species to achieve this seemingly utopian vision. We've entered into the final minutes of the eleventh hour, and our midnight transformation is upon us. The time left to institute conscious positive change is running precariously near to the possible end of human existence. We each need to consider all of our actions in light of the fact that we are dangerously close to destroying the only planet we have available to us. Then, we need to "be the change" we want to see manifest globally.

Now is the time and we are the people. It's up to you and me and our neighbors a well as all the crewmembers on board Spaceship Earth. As Bucky so often reminded his audiences, *"The cosmic question has been asked, 'Are humans a worthwhile to Universe experiment?'"* I say "YES," and the only way we can survive and thrive is if each of us does our piece and shares our gifts to create the vision that Bucky first articulated as ***"a world that works for everyone."***

# Index of Quotations and Guest Commentators

**1.5** "We now have the resources, technology and know-how to make of this world a 100% physical success."
*Utopia or Oblivion*

BY BARBARA MARX HUBBARD

**1.6** "Love is metaphysical gravity."
*Critical Path*

BY GARY ZUKAV

**1.7** "Nature always knows what to do when it takes over after humans have signed off. Nature never vacillates in its instantaneous decisions."
*Critical Path*

BY JACK ELIAS

UNIVERSAL PERSPECTIVE OF GRATITUDE AND EQUANIMITY
### *Chapter Conclusion*

**1.8** "I don't have any favorite places or people. I love the whole show. A large number of beautiful people have taught me a great deal, and I am deeply indebted to them for their support."
*Fuller to Kay Elliot, Washington Star (September 1975)*

## CHAPTER 2

PURPOSE: HUMANS ON EARTH

**2.1** "The most important thing about Spaceship Earth—an instruction book didn't come with it."
*Operating Manual for Spaceship Earth*

**2.2** "The most special thing about me is that I am an average man."
*Fuller to J. Maxwell Smith, Jr. (March 1974)*

BY ROBERT WHITE

**2.3** "It is not for me to change you. The question is, how can I be of service to you without diminishing your degrees of freedom?"
*The Future of Business (August 1981)*

BY GREG VOISEN

**2.4** "Humanity is taking its final examination. We have come to an extraordinary moment when it doesn't have to be you or me. There is enough for all. We need not operate competitively any longer. If we succeed, it will be because of youth, truth, and love."
*Buckminster Fuller: An Autobiographical Monologue/Scenario*

BY HAZEL HENDERSON

BY T. J. MACKEY

**2.5** "All children are born geniuses. 999 out of every 1,000 are swiftly and inadvertently degeniused by adults."
*Draft Preface for Mrs. John S. Lillard (August 1975)*

BY ANNA BESHLIAN

BY STEPHEN GARRETT

**2.6** "I live on Earth at present, and I don't know what I am. I know that I am not a category. I am not a thing—a noun. *I seem to be a verb*, an evolutionary process—an integral function of Universe."
*I Seem to Be a Verb*

BY VELCROW RIPPER

BY THOMAS MYERS

Purpose Without Control
***Chapter* Conclusion**

**2.7** "Anyone who thinks that humans on this Earth are running Universe or that Universe was created only to amuse or displease or bore humans is obviously ignorant."
*How To Make Our World Work, New Dimensions Lecture Series (San Francisco)*


**Chapter 3**

Being: Showing Up Fully

**3.1** "Dare to be naïve."
*Synergetics: Explorations in the Geometry of Thinking*

**3.2** "You have to decide whether you want to make money or to make sense because the two are mutually exclusive."
*Technology: Enchantment vs. Disenchantment (June 1977)*

By Randolph L. Craft

**3.3** "So I vowed to keep myself alive, but only if I would never use me again for just me—each one of us is born of two, and we really belong to each other. I vowed to do my own thinking, instead of trying to accommodate everyone else's opinion, credos, and theories. I vowed to apply my own inventory of experiences to the solving of problems that affect everyone aboard planet Earth."
*Critical Path*

By Dr. Joel Levey

**3.4** "Never mind if people don't understand you, so long as no one misunderstands you."
*Buckminster Fuller: At Home In Universe*

By James Roswell Quinn

**3.5** "There is something patently insane about all the typewriters sleeping with all the beautiful plumbing in the beautiful office buildings—and all the people sleeping in the slums."
*DSI Press Conference (June 1972)*

**3.6** "Unconscious decisions have consequences. Our assumptions drive our priorities, and in many cases we don't even acknowledge they are there. Innovation arises from questioning the old assumptions."

BY STEPHAN A. SCHWARTZ

**3.7** "We are called to be architects of the future, not its victims. The challenge is to make the world work for 100% of humanity in the shortest possible time, with spontaneous cooperation and without ecological damage or disadvantage of anyone."
*The Future of Business (August 1981)*

BY DC CORDOVA

**3.8** "I have learned that it is possible to stand and think out loud from the advantage of our most effective possible preparation, which is all recorded, and on tap in our brains and minds. Advance thought about our discourse spoils it. There, awaiting its anytime employment by our brain-scanning mind, is the ever recorded and highlighted inventory of our life-long experiences integrated with all the relevant experiences others have communicated to us. Out of this inventory, your live presence catalyzes my freshly reconsidering thoughts relevant to our mutual interests."
*Environment and Change* (Originally deleted from *Operation Manual for Spaceship Earth*)

COMING TOGETHER
*Chapter Conclusion*

**3.9** "If two of us meet and you take a paper out of your
pocket and start reading a speech, I will say, 'Let me
have that. I can read it myself more effectively.' I am
confident that live meetings catalyze swift awareness of
the particular experiences of mutual interest regarding
which our thoughts are spontaneously formulated. Live
meetings often become pivotal in our lives."
*Environment and Change (Originally deleted from Operation*
*Manual for Spaceship Earth)*

**CHAPTER 4**

DESIGN: TRIMTAB ON THE PATH

**4.1** "The greatest of all faculties is the ability of the
imagination to formulate conceptually. Artists have
kept the integrity of childhood alive until humanity
reaches the bridge between the arts and sciences."
*Fuller: Who Will Man Spaceship Earth? (September 1971)*

**4.2** "Environment is stronger than will."
*The Future of Business (August 1981)*

BY BOBBI DEPORTER

**4.3** "Revolution by design and invention is the only
revolution tolerable to all men, all societies, and all
political systems anywhere."
*Oregon Lecture #2 (July 1962)*

**4.4** "We are powerfully imprisoned in these Dark Ages
simply by the terms in which we have been conditioned
to think."
*Cosmography*

BY ZOE WEIL

**4.5** "The most important thing to teach your children is that the Sun does not rise and set. It is the Earth that revolves around the Sun. Then teach them the concepts of North, South, East, and West, and that they relate to where they happen to be on the planet's surface at that time. Everything else will follow."
*WNBC-TV Interview, 1983*

BY ANN MEDLOCK

**4.6** "Faith is much better than belief. Belief is when someone else does the thinking."

BY SATYEN RAJA

**4.7** "How often I found where I should be going only by setting out for somewhere else."
*How To Make Our World Work, New Dimensions Lecture Series (San Francisco)*

BY ROSHI JOAN HALIFAX

BY LD THOMPSON

START WITH UNIVERSE—END UP WITH CONTRIBUTION
*Chapter Conclusion*

**4.8** "I didn't set out to design a house that hung from a pole, or to manufacture a new type of automobile, invent a new system of map projection, develop geodesic domes, or Energetic-Synergetic geometry. I started with the Universe—as an organization of energy systems of which all our experiences and possible experiences are only local instances the principles operating in Universe, I could have ended up with a pair of flying slippers."
*Encyclopedia Britannica article by Robert W. Marks (August 1974)*

## Chapter 5

## CONSCIOUSNESS; Waking Up To Being Awake

**5.1** "There is nothing in a caterpillar that tells you it's going to be a butterfly."
*How To Make Our World Work,* New Dimensions Lecture Series (San Francisco)

**5.2** "Very very slow changes humans identify as inanimate. Slow change of pattern they call animate and natural. Fast changes they call explosive, and faster events than that humans cannot see directly."
*Electromagnetic Spectrum (Oct. 1970)*

**5.3** "Our children and our grandchildren are our elders in universe time. They are born into a more complex, more evolved universe than we can experience or than we can know. It is our privilege to see that new world through their eyes."

*By John Robbins*

*By Ocean Robbins*

*By Lynne Twist*

*By Marilyn Schlitz*

**5.4** "No human can prove, upon awakening, that they are the person who they think went to bed the night before or that anything they recollect is anything other than a convincing dream."
*Cosmic Fishing*

**5.5** "If the success or failure of this planet, and of human beings, depended on how I am and what I do, how would I be? What would I do?"
*How To Make Our World Work, New Dimensions Lecture Series (San Francisco)*

**6.4** "There are no 'good' or 'bad' people, no matter how offensive or eccentric to society they may seem. You and I didn't design people. God designed people. What I am trying to do is to discover why God included humans in Universe."
*Critical Path*

By Jamal Rahman

**6.5** "Think about a stadium of seventy-five thousand people. It's really quite a large crowd, isn't it? Every day seventy-five thousand people die of starvation despite the fact that we have plenty of food for everyone. Our distribution systems, our nations, all the different kind of separateness blocks the whole thing. Just think of it. Simply because we're badly organized, we're not taking care of it."
*Oregon Lecture #5 (July 1962)*

**6.6** "If humanity does not opt for integrity we are through completely. It is absolutely touch and go. Each one of us could make the difference."
*Only Integrity Is Going To Count (1983)*

By Dr. David Gruder

Making Our Spaceship Work
## *Chapter Conclusion*

**6.7** "We are on a spaceship; a beautiful one. It took billions of years to develop. We're not going to get another. Now, how do we make this spaceship work?"
*Critical Path*

## CHAPTER 7

## ACTION: MAKING A DIFFERENCE

**7.1** "Take the initiative. Go to work, and above all co-operate and don't hold back on one another or try to gain at the expense of another. Any success in such lopsidedness will be increasingly short-lived."
*Operating Manual for Spaceship Earth*

**7.2** "There is no use talking about bright ideas. Everyone has bright ideas. There is no use talking about an artifact until we reduce it to practice, until we see whether Nature permits it, whether society permits it."
*Critical Path*

*By Dr. Cherie Clark*

**7.3** "What can a little man effect toward such realizations in the face of the formidable power of great corporations, great states, and all their know-how, guns, monies, armies, tools and information?

The individual can take initiatives without anybody's permission. Only individuals can think, and can look for the principles manifest in their experiences that others may be overlooking because they are too preoccupied with how to please some boss or with how to earn money, how to take care of today's bills. Only the individual disregards his fears and commits himself exclusively to reforming the human environment by developing tools that deal more effectively and economically with evolutionary challenges.

Humans can participate—consciously and competently—in fundamental ways, to changes that are

more favorable to human life. It became evident that the individual was the only one that could deliberately find the time to think in a cosmically adequate manner."
*Buckminster Fuller: An Autobiographical Monologue/Scenario*

By Peter Meisen

**7.4** "My working assumption is that we are here as local Universe information gatherers. We are given access to the divine design principles so that from them we can invent objectively the instruments and tools that qualify us as local Universe problem solvers in support of the integrity of an eternally regenerative Universe."
*Only Integrity Is Going To Count (1983)*

By Michelle Levey

**7.5** "You never change things by fighting the existing reality. To change something, build a new model that makes the existing model obsolete."
*Critical Path*

By Bill Kauth

By Rick Ingrasci

By Michela Miller

**7.6** "The best way to predict the future is to invent it."

By Hunter Lovins

Take Action and Do Something!
**Chapter Conclusion**

**7.7** "We are not going to be able to operate our Spaceship Earth successfully nor for much longer unless we see it as a whole spaceship and our fate as common. It has to be everybody or nobody."
*Only Integrity Is Going To Count (1983)*

# SOURCES OF THE QUOTATIONS

During the decades I have studied Buckminster Fuller's life, wisdom, and work, I discovered that his seemingly complex and diverse message, like Nature itself, is actually quite simple. Although he lived by the adage *"Don't try to make me consistent. I am learning all the time,"* the essence of his communications to fellow travelers on board Spaceship Earth was consistent.

From his first publishing of *4D* magazine in 1927 through his final public "thinking out loud" lecture in 1983, his guidance followed a constant theme—Nature is attempting to make humans a success, and we now have the capacity to be cooperative partners in that endeavor if we first learn to collaborate with and be supportive of our fellow human beings.

Bucky tended to recount this and his other guiding messages again and again at each of his lectures and in his writings, but the exact words he used would often be different. I have found a quotation written or spoken in many different forms. They may not contain the exact same words, but their essence and intention are constant.

Bucky did his best to be of service to us all by making his thoughts and ideas as publicly available as possible. He did not have any organization disseminating or monitoring the information that he felt was so vital to the success of humankind.

Because of these reasons, the quotations listed in this book can be found in many different forms and sources. The source for each quote that is listed in this Index is usually one of many, and I do not claim that it is the only or best place that the quotation can be found.

Several people have worked diligently to verify that each quotation was said or written by Buckminster Fuller, however we have not been able to do so for every quote. Some of the wisdom in this book was heard by the commentators or written down by individuals who unofficially took them to heart and shared them with others. Thus, no formal record of their source exists. Like the insights of the great master teachers, many of Bucky's words were written down after the fact to be shared by people who were in his presence and felt responsible to capture this wisdom as best they could.

If anyone is able to substantiate the source of any unlabelled quotation in this book please notify us, and we will publish that source in future editions. Thank you for your generous support in this endeavor.

L. Steven Sieden. July 12, 2011. My personal email address is ssieden@gmail.com.

# Acknowledgments

Nothing as daunting as creating this book could be accomplished as a solo effort. Even though I often went for days without speaking to another person while writing and editing *A Fuller View*, it is the result of a genuine collaboration. Most notable are the forty-two generous souls who said "yes" to my Guest Commentator invitation. Without the kind and generous support of these people, the perspective of *A Fuller View* would be far from fuller. These people are …

*Gary Zukav, John Robbins, Lynne Twist, Marilyn Schlitz, Barbara Marx Hubbard, Werner Erhard, Hunter Lovins, Hazel Henderson, Roshi Joan Halifax, Ocean Robbins, Bill Kauth, Jack Elias, Stephen Garrett, James Roswell Quinn, Michelle Levey, Dr. Joel Levey, Kevin Todeschi, TJ McKay, Anna Beshlian, DC Cordova, David Spangler, Jim Reger, David Irvine, Michela Miller, Dr. David Gruder, Justine Willis Toms, Bobbi DePorter, Ann Medlock, LD Thompson, Zoe Weil, Peter Meisen, Robert White, Velcrow Ripper, Satyen Raja, Jamal Rahman, David McConville, Greg Voisen, Thomas Meyers, Randolph L. Craft, Dr. Cherie Clark, Lisa Matheson, Dr. Rick Ingrasci, Stephan Schwartz and Michael Wiese.*

My confidence in the necessity of this book continued to grow as the list of Guest Commentators expanded like a snowball rolling downhill. And without the initial push from my friends Bill Kauth, Michelle Levey, and Joel Levey that ball would not have rolled so smoothly or quickly.

The other person to whom this book and I owe so much is Michael Wiese, a publisher all authors should have the privilege of working with and a genuine *mensch*. Michael and I cooked up this idea via email late one winter night (it was actually daytime in England, where he lives), and he has been a true partner throughout this process. His associate Manny Otto has also been kind and generous well beyond what I have ever experienced dealing with publishers and editors. Their copy editor Andrew Beierle has made the book even more impeccable (as Bucky would have wanted it), and he did his work with a grace and ease that reduced my stress.

I've felt that this book was destined to be since August 31, 2009, when I came about as close to dying as possible. On that day I was gifted with "extra" years of life as a result of quadruple bypass open-heart surgery (I call it my "open heart initiation"), and I had a sense that there was a clear reason for my survival.

Three people were critical to me making it through that initiation. Without Laurie Ross, Jill Clarice Sieden, and Mark Blair, this book would never have been written because I would have died. That incident and the months of recovery caused me to remember Bucky's admonition that we each have unique gifts to contribute. It also led me to realize that I was the person with the experience and skill to share a unique vision of Bucky's legacy through writing and presentations. Just as the voice spoke to him and told him that he was a unique link in a chain that he did not have a right to break, my heart incident caused me to assess my situation and find what needed to be done that was not being attended to.

Still, without the generous support of others, this book would never have come to fruition. Key among those who have sustained me along this path are people I consider master teachers who have played a significant role in my life since I took my first meditation/yoga class in 1975. These amazing people are too numerous to mention or recall. Still, I would like to acknowledge the Seventh Dzogchen Ponlop Rinpoche, Joel and Michelle Levey, Mother Mary, Jean Houston, Alexander Everett, Werner Erhard, Alberto Villoldo, Elizabeth Cogburn,

Swami Veda Bharati, Mary Adams, Hank Wesselman, Marshall Thurber, Ipupiara, Michael Hague, and Pema Chodron.

And no acknowledgement of Bucky Fuller's work and wisdom would be complete without thanking the family of Buckminster Fuller—especially his daughter Allegra Fuller Snyder and grandson Jaime Synder—for keeping his legacy alive through the Buckminster Fuller Institute. To them, the BFI (www.bfi.org) and its president David McConville, "we the people" will always be grateful.

△ △ △

Many others, too numerous to recall or remember, have supported me in this journey. As Bucky once wrote, *"A large number of beautiful people have taught me a great deal, and I am deeply indebted to them for their support."* The following is a list of all those I recall at this moment. I apologize in advance to anyone I forgot in my haste to complete this project. I am grateful to you all for your kind and generous support.

Wayne Greenfield (www.WayneGreenfield.com), Tauri Senn and Michael Gamble all of whom caught many mistakes in the manuscript, ManKind Project Brothers, the Nalandabodhi Seattle Sangha—especially our Monday night Dharma Sangha, Diana Elmer, Elizabeth Kawakami, Steve Heck, Kelly Clifton-Purdy, Dr. Tim Clanton, Dr. Dirk Farrell, Jack Elias, Hope Maltz, Alex LaVilla, Kathleen Carrara, Dr. Bruce Beaulaurier , Stanley Sabre, Dr. Dean Chier, Susan Gray, Michael Hartzell, Jim Burbidge, Shefali Sinha, Karolyn McKinley, JR Gillespie, Cheryl Valk, Malcolm McKay, Paul R. Scheele, Andy Bradford, Linda Francis, Lisa Kennedy, Cassidy Fritz, Elizabeth Mattis-Namgyel, Lori Weismann, Sally King, Auben Pamela Gail MacKay, Dr. Sheila Dunn-Merritt, Dr. Jason MacLurg, Robert Greczanik, Diane Brooks, Dr. Jeffrey A. Hirsch, Greg Voisen, Lovorka Knezevic, Ken

Lee, Don Ross, Ceci Miller, Joel Levey, Michelle Levey, Mike O'Shea, Vicki Robin, Jon Witte, Elle McSherry, Trey Styles, Mike Dooley, Deborah Drake, Phil Sieden, Sherwin Sieden, Michael Sieden, Irving Sieden, Sam Sieden, Louise Lent Sieden, Rex Maruca, Dr. Gabriel Aldea, Dr. Rubin Maidan, Alice Villa, Lynn Conrad Marvet, Ross Quinn, Greg Torvik, Maureen Goldish, Todd Bernstein, Sheila Baker, Mark Power, Robert Lee, Nadine Selden, Stuart Horn, Hobbs Sieden, David Sieden, Karley Sieden Sweet, Daniel Sieden, Jesse Sieden, Frank Dollar, and all the people who smile at me during my daily walk or jog around Green Lake.

May all these generous people continually experience the bounty of our abundant Spaceship Earth and the gratitude of all those who benefit from this book.

# Epilogue – Dedication

In the Buddhist tradition, we always complete a meditation, teaching, or anything of value with a "dedication of merit." In the publishing tradition, the dedication is at the beginning of the book, and I followed that publishing tradition at the front of this book.

I complete this portion of what I have often described as my plunge into the depths of the "Bucky Dharma" with the following dedication. I do this understanding that there are no beginnings or ending in absolute, universal reality and knowing that everything we do is with and for all beings.

*May all beings be happy and peaceful.*
*May all beings be free from fear and pain.*
*May all beings be healthy and strong.*
*May all beings live with love and compassion.*
*May all beings fully awaken and be free.*

To the best of his ability Dr. R. Buckminster Fuller lived a happy, healthy, peaceful life free from fear and pain with a great strength of love and compassion, and these attributes allowed him to awaken and be free more fully than most of his peers. He also fully shared all that he could of what he had learned on his journey.

This small volume represents a tiny fraction of the wisdom and insights he uncovered and of the seeds that he sowed and of the love he gave so freely. Please take whatever you find relevant and true to your experience. Use it and share it as often as possible. That's how we create sufficiency and abundance, and how we manifest our true mission as compassionate, mindful stewards and navigators creating *"a world that works for everyone"* on board a tiny, fragile planet that Bucky called Spaceship Earth.

L. Steven Sieden—Passenger and Crewmember
July 12, 2011
Seattle, Washington, Spaceship Earth

# ABOUT THE AUTHOR

L. STEVEN SIEDEN has been a student of and advocate for Dr. R. Buckminster Fuller since 1981 when Steven spent three months on a beach studying *Critical Path*. He worked on Fuller's Integrity Days—the last series of public appearances Fuller made, and when Fuller died in July 1983, Steven began working with The Buckminster Fuller Institute to produce events using a portion of Fuller's vast recorded archive. He also began learning more about Fuller's life and ideas, and his research led to the 1988 biography, *Buckminster Fuller's Universe* (currently Basic Books 2000).

Since 1981, Steven has continued to study and apply Fuller's wisdom to his own life while writing and lecturing on the lessons we can all learn from Bucky. Following a major "heart initiation" (a.k.a. quadruple bypass open heart surgery), Steven had his own epiphany and got the message that Fuller's ideas needed to be spread more fully, especially among the younger generations. Thus, the impetus for *A Fuller View* and the volumes to follow.

Steven lives in Seattle, Washington, on planet Earth where he continues to receive "downloads" of Buckminster Fuller insights and support while connecting with and supporting others who have access to Fuller's wisdom and experience. For more information go to www.BuckyFullerNow.com or www.BuckminsterFullerNow.com.

Author photo by Lori Davis-Sandoval, www.LakeUnionRecording.com

If success or failure
of this planet and of human
beings depended on
how I am and what I do

How would I be?
# What would I do?
— R. Buckminster Fuller

## THE
## BUCKMINSTER
## FULLER
## CHALLENGE

The Buckminster Fuller Challenge is an annual international design Challenge awarding $100,000 to support the development and implementation of a strategy that has significant potential to solve humanity's most pressing problems. Named "Socially-Responsible Design's Highest Award" by Metropolis Magazine, it attracts bold, visionary, tangible initiatives focused on a well-defined need of critical importance. Winning solutions are regionally specific yet globally applicable and present a truly comprehensive, anticipatory, integrated approach to solving the world's complex problems.

## CHALLENGE.BFI.ORG

The Buckminster Fuller Institute I 181 N 11th St I Suite 402 I Brooklyn, NY 11211
T: 718 290 9283 I F: 718 290 9281 I challenge@bfi.org

an imprint of **MICHAEL WIESE PRODUCTIONS**

# DIVINE
ARTS

DIVINE ARTS sprang to life fully formed as an intention to bring spiritual practice into daily living. Human beings are far more than the one-dimensional creatures perceived by most of humanity and held static in consensus reality. There is a deep and vast body of knowledge — both ancient and emerging — that informs and gives us the understanding, through direct experience, that we are magnificent creatures occupying many dimensions with untold powers and connectedness to all that is. Divine Arts books and films explore these realms, powers and teachings through inspiring, informative and empowering works by pioneers, artists and great teachers from all the wisdom traditions.

We invite your participation and look forward to learning how we may better serve you.

Onward and upward,

Michael Wiese
Publisher/Filmmaker

DivineArtsMedia.com